Reinventing Innovation

Aaron C.T. Smith · Fiona Sutherland
David H. Gilbert

Reinventing Innovation

Designing the Dual Organization

Aaron C.T. Smith
RMIT University
Melbourne
Australia

David H. Gilbert
RMIT University
Melbourne
Australia

Fiona Sutherland
La Trobe University
Melbourne
Australia

ISBN 978-3-319-57212-3   ISBN 978-3-319-57213-0   (eBook)
DOI 10.1007/978-3-319-57213-0

Library of Congress Control Number: 2017938546

Cover Illustration: © saulgranda/Getty

Printed on acid-free paper

This Palgrave Macmillan imprint is published by Springer Nature
The registered company is Springer International Publishing AG
The registered company address is: Gewerbestrasse 11, 6330 Cham, Switzerland

CONTENTS

LIST OF FIGURES

LIST OF TABLES

CHAPTER 1

The Innovation Imperative

Abstract This chapter establishes the context, background, and aims for the book noting that organizations must manage ostensibly opposing forms, such as stability and change, and freedom and accountability. Organizations therefore need to manage simultaneously for both efficiency (exploitation) and flexibility (exploration). The chapter foreshadows the book's key premise that the exploration—exploitation tension represents a 'duality' that must be embraced rather than resolved. It foreshadows the results of several longitudinal research cases demonstrating that innovation works best when in concert with efficiency, rather than as a stand-alone or as an alternating on and off priority. Finally, the chapter introduces how innovation capacity can be developed through ambidexterity capability.

Keywords Duality · Explore · Exploit · Innovation

INTRODUCTION

When faced with unprecedented competition and adversity, organizations tend to fall back upon what they have always done best. But what happens when exploiting existing strengths is no longer good enough? In conditions where exploring new opportunities has become essential, the current challenge demands the pursuit of both exploitation *and* exploration at the same time, that is, both efficiency and innovation. Finding a way to navigate through the seemingly incompatible tensions between exploration and exploitation—or innovation and control—has become the quintessential problem characterizing management in the technologically furious, global business world of the twenty-first century. Yet, maintaining an efficient return on current operations while also developing a burgeoning pipeline of new products remains elusive.

Based on research findings, *Reinventing Innovation* charts the new innovation imperative, where exploitation and exploration must flourish at the same time. However, these two seemingly contradictory objectives cannot be achieved through some bland compromise, or by switching priorities back and forth. Only a 'dual' organization capable of amplifying the tension can optimize efficiency while seeding innovation. In this book, we examine dual organizing with particular attention on how to build 'dualities' in organizations. Our approach draws on theory and research, but it also makes heavy use of in-depth cases based on our own longitudinal studies. By bringing the theory, research, and case lessons together, we aim to expose the foundations of dual organizations, or what we refer to as 'ambidexterity capability.' We employ the term in reference to an organization's ability to perform explore and exploit tasks at the same time. However, while all organizations can do at least a little of each at any given time, ambidexterity capability means that they can deliver on both in significant, sustainable amounts.

Armed with ambidexterity capability, dual organizations can meet the twin mandate of current and future performance. Our investigation of dual organizing emphasizes the simultaneous escalation of both explore and exploit ventures in order to sustain a productive tension where creativity coexists with control. Unlike approaches focusing exclusively on innovation, which propose independent innovation programs and processes or recommend a balance between predictable maintenance and risky innovation, here we advocate for more of both. In the resulting tension, opportunities emerge for organizations to assimilate ambiguity, while still sustaining an orderly return on current strengths.

The dual organizing story presented in the forthcoming chapters advances a new perspective on the exploration-exploitation problem. It further examines duality theory's application through original and longitudinal empirical case studies, subsequently converting the theory and data into practical recommendations for implementation. These propositions revolve around the day-to-day challenge of developing ambidexterity capability.

What might be called the explore—exploit problem constitutes the contemporary incarnation of the 'innovator's dilemma.' Organizational leaders need to find new ways to accommodate the structural and managerial tensions that accompany a commitment to innovation through both exploitation ('tight' structures, control, continuity, stability, conventional reporting and performance measures) and exploration ('loose' structures, flexible, responsive, experimental, and evolving). Numerous theories and principles inform methods for introducing either control/quality programs or innovation/change programs. However, a theoretical and practical framework to integrate both in high doses remains underdeveloped. In this book, we offer a theoretical, empirical, and practical background for students, managers, and leaders seeking to improve their responses to the universal problem of innovating for the future while delivering results in the present.

The call for organizations to be agile in exploiting the core business of today, while simultaneously exploring the business of tomorrow, has been prominent for many decades. However, no comprehensive theory has emerged to guide organizations toward authentic ambidexterity, where an organization achieves success in both organizing forms simultaneously. Instead, various contingent forms of ambidexterity have been advocated, such as temporal switching between explore and exploit. Such 'ambidextrous sequencing'—switching on and off between an exploitative framework (premised by efficiency, stability, and control) and an explorative framework (invoking innovation, adaptability, and risk-taking)—represents a significant challenge for leaders and managers trapped between shifting priorities. This book is motivated by the need to respond to the pervasive, enduring, and ubiquitous problem of ambidextrous organizing to solve the explore—exploit problem. To be successful, organizations simply must be great at both.

This chapter establishes the context, background, and aims for the book. It begins by noting that organizations must manage ostensibly opposing forms, such as stability and change, and freedom and accountability. In short, organizations need to manage simultaneously for both efficiency (exploitation) and flexibility (exploration).

The chapter foreshadows the book's key premise that the exploration—exploitation tension represents a 'dual' tension that must be embraced rather than resolved. Supporting this argument, the chapter outlines the results of our longitudinal research around the explore—exploit problem. Our foundational suggestion is that innovation works in concert with efficiency where the two peak together, rather than as a stand-alone or as an alternating on and off priority. From this base principle, we address the mechanics of success through real-world examples of how ambidexterity capability has been built.

We think that much effort and resources are wasted on directionless 'innovation' and on the prosecution of innovation at all costs even when maximizing returns from existing products is warranted. In addition, we maintain that many innovation programs declare ambitions of novelty and revolution when a better focal point would advance some 'innovative' progress around products that are well known to be in demand. Disruptive innovation is not the only innovation. It need not be measured by the de-installation of what is already working and replaced by something else that does much the same sort of thing. Hence, there is an endless stream of innovation programs that get nowhere, instead of an overarching and sustainable program that allows an interface between sound returns on existing investments and pipeline ambitions. The chapter also introduces the forthcoming chapters and their contents. We continue next with a brief introduction to the pivotal contextual and historical issues that presaged the need for dual organizations and ambidexterity capability.

THE ORGANIZING QUEST

Business commentators have been talking about the 'new' organization in various forms since the 1950s. The prolific resulting literature on the so-called new forms of organizing, while divergent in approach and emphasis, has embodied a shared rhetorical message: 'the world is changing, traditional bureaucracy is bankrupt and the future is now—or at least soon' (Nohria and Berkley 1994, p. 108). Typically, the need for change features prominently in justifications for organizing differently. As a result, industry deregulation, free-trade markets, and the rise of Asia and its subcontinent, for example, have all been used to explain a newfound impetus for global competition, connectivity, and economic activity. At the same time, disruptive technological advances—including cloud-based commerce, the Internet of Things, and machine-based learning through artificial intelligence—mean that borders and geography no longer impede the transfer of information, capital, people, or products. This new 'Information Age' has occurred alongside a gradual, but inexorable shift in the focus of developed economies from the manufacturing and commodities sectors to the information, communications, and services sectors. A technology revolution in tandem with rising consumer expectations and access to global markets has also contributed to shorter product life cycles and the need for go-to-market speed and flexibility. In many industries, technological advances have eroded the traditional barriers to entry where high start-up costs and economies of scale no longer present severe obstacles. We have therefore seen an explosion of fast, small, and nimble companies entering markets that were once the protected sanctum of industry behemoths.

A review of the commentary about all of this environmental complexity and uncertainty reveals a narrative about the need for urgent change toward more flexible forms of working. For example, organizations have been urged to downsize, decentralize, de-bureaucratize, decouple, differentiate, empower, innovate, integrate, and involve. In an apocalyptic style, labels abound for the new organizational forms that will save the day: boundary-less, network, platform, virtual, clickable, hybrid, modular, horizontal, shamrock, loosely coupled, individualized, learning, knowledge-based, and cellular (DiMaggio 2001; Schilling and Steensma 2001; Palmer and Hardy 2000; Whittington et al. 1999). At the core was a call for a horizontal, boundaryless organization—a network of interconnecting parts comprising smaller, cross-functional business units with inter-organizational partnerships and alliances (Ghoshal and Bartlett 1995; Dunphy and Stace 1993; Limerick and Cunnington 1993; Ulrich and Wiersema 1989). In short, as the world was changing so too were organizations urged to match that change. The answer, it seemed, was to shrug off the oppressive, bureaucracy-laden shackles of traditional command and control systems, and experiment with more flexible and agile organizational forms encouraging innovation, exploration, and self-learning (Bahrami 1992; Jonsson 2000).

If the answer in the new, knowledge-driven business world was to be found in organizing forms designed for swift adaptation to change, the 'old' bureaucratic model had to be tossed aside. Adherence to hierarchy, stability, uniformity, and specialization, designed to exert authority and control over a largely uneducated workforce, all had to go. Indeed, if organizations hoped to survive and succeed in complex, high-velocity (Eisenhardt and Brown 1998), chaotic (Dijksterhuis et al. 1999) environments, they had to make flexibility the focus (Schilling and Steensma 2001). Moreover, organizing structure was seen as the critical feature in continually transforming to accommodate the needs of the competitive environment (Rindova and Kotha 2001).

Meanwhile, claims proliferated that organizations needed new ideas, paradigms, and practices in order to cope with the unprecedented demands that the global, knowledge economy had delivered (e.g., Kelly 1998). For example, by the late 1990s, the key themes in business literature could be summarized under seven labels: technology, globalization, competition, change, speed, complexity, and paradox (Tetenbaum 1998). In response, management theorists started to think of organizations as robust and dynamic systems characterized by a constructive tension between order (the push of exploitation) and disorder (the pull of exploration). It was a description that gained traction in the management world—at least for consultants—under the 'edge of chaos' nomenclature (Lewin et al. 1999). As the edge of chaos metaphor gathered momentum, organizational theorists and practitioners slowly relinquished a linear, dualism view in favor of a less neat, duality view.

Dualism is a long-enduring approach to classification where the object of study is divided into paired and opposite elements, the most common examples being mind/body and theory/practice (Jackson 1999). Duality theory on the other hand—a by-product of Gidden's (1984) structuration theory—suggests that dualism elements may be independent and conceptually distinct, rather than opposed (Smith and Graetz 2006). Thus, theorists who employ duality theory 'can maintain conceptual distinctions without being committed to a rigid antagonism or separation of the two elements being distinguished' (Jackson 1999, p. 549).

In practice, a duality mindset translates into organizational designs built around seemingly opposing forces such as efficiency and innovation, hierarchy and networks, global operating control and local responsiveness, and centralized vision and decentralized autonomy (Child and McGrath 2001; Pettigrew et al. 2003). Yet, a dualities-aware perspective does not favor one side of the organizing duality pole over the other. Rather, it recognizes that both have merit, thus encouraging a creative tension between opposing forces such as a short- and long-term focus, differentiation and integration, external and internal orientation, and continuity and change (Evans 1999). It is at this point of dynamic tension, 'when organisations are neither so structured that change cannot occur, nor so unstructured that chaos prevails' (Evans 1999, p. 335), when both efficiency and innovation are most likely to take place (Lewin et al. 1999). Although perhaps a little counterintuitive, success in the new world of business meant that organizations had to balance on the cusp between order and disorder (Pettigrew and Fenton 2000).

NEW FORMS, CHANGE, AND INNOVATION

As environmental turbulence escalated, organizing forms became increasingly associated with organizational change and innovation. At first, management practitioners reacted to external uncertainty with tentative and contained experiments in restructuring (Dijksterhuis et al. 1999). Mostly, this came through a cautious weakening of rigid bureaucracies in the hopes of stimulating more flexibility and responsiveness. Slowly, the intuitive hesitancy to relinquish control was overtaken by more aggressive forays into flatter structures, devolved decision-making responsibility, improved intra-organizational collaboration, partnerships with other organizations, participative and empowering management, and a generally more creative orientation (DiMaggio 2001; Pettigrew et al. 2003; Dijksterhuis et al. 1999; Volberda 1998). In short, traditional forms of organizing exemplified by the bureaucracy were considered obsolete in a world where change was the sovereign currency. Vigorous calls prescribing the flattening of hierarchies, horizontal collaboration, diminished formalization, and a weakening of ties between workers and firms (DiMaggio 2001) had become the convention. But it was not to be a smooth exchange of one for the other. New forms adherents claimed that network-driven, decentralized, malleable structures worked better in fluid environments than traditional, hierarchical, rule-centered bureaucracies. However, it turned out that there was a catch.

Evidence from organizations that had attempted to introduce new forms of organizing showed that the wholesale replacement of traditional forms for new forms was wildly unrealistic (Fenton and Pettigrew 2000; Pettigrew et al. 2003). Despite the claims, in practice bureaucracy is robust and resilient because it contains safeguards through line management, job specificity, and control systems (DiMaggio 2001). Of course, the very nature of redundancy makes flexibility troublesome, particularly in a changeable marketplace. On the other hand, flexibility in the new forms costume leaves everything to the chance whim of empowered employees and teams. As organizational researchers like us discovered, this choice can be unpalatable or even untenable (Graetz and Smith 2009; Palmer and Dunford 2002). A consequence of this non-choice was the concomitant use of both traditional and new forms in a kind of paradoxical and uneasy partnership described as a duality (Pettigrew and Fenton 2000; Whittington et al. 1999). New forms bolstered rather than replaced traditional forms of organizing.

Contradictory though it may seem in principle, the simultaneous presence of the two forms of organizing has not been a failure (O'Reilly and Tushman 2004; Palmer and Dunford 2002; Raynor and Bower 2001; Volberda 1998). Research focusing on the assumed countervailing tendencies between efficiency and effectiveness—new forms and traditional forms—revealed that the dilemma for managers is not only real and paradoxical, but also healthy and normal (Adler et al. 1999). The relationship between the two forms is probably better understood in terms of complementary approaches that flex in tension, rather than contradictory forces that collide in opposition (Sanchez-Runde and Pettigrew 2003).

Dualities represent an alternative interpretation on the interplay between conventional philosophical adversaries. It is probably useful to keep in mind that the focus is on 'forms of organizing, not forms of organization' (Quinn et al. 1998, p. 162). Our position maintains that dualities offer a nascent but high-potential theoretical platform from which to consider concepts bound at either end of the organizing forms continuum. Most particularly, dualities provide the theoretical substance from which we can test some ideas in practice. Our approach has been to spend years working with and observing organizations that have pursued both explore and exploit objectives. Accordingly, our findings are reproduced in the forthcoming chapters, with the lessons they display stimulating a range of recommendations for real-world application. Chiefly, we argue for structures and practices that advance ambidexterity capability.

On the upside, despite the early fervor and subsequent failure of introducing innovative forms of organizing while abandoning traditional forms, our research has revealed that high-performing companies have taken a third path involving the seemingly contradictory but simultaneous combination of the two (Whittington et al. 1999).

Dualities and the Explore—Exploit Dilemma

Innovation is surely an advantageous reaction to market change, but it does however, demand the stability of bureaucratic order in order to come about in the first instance. It might also be worth remembering that not all organizational activity reacts favorably under edge of chaos conditions, such as legal obligations, employee remuneration, and risk management. Our argument is that dualities offer a starting point for working through the explore—exploit collision. By way of introduction, we offer a handful of guiding principles or characteristics of dualities to serve as a foundation leading to the next chapters. Keep in mind that dualities occur when high levels of both exploit (control) and explore (flexibility) organizing forms exist at the same time.

First, dualities assume that traditional and new forms of organizing are compatible in the sense that they can coexist in a kind of productive tension (Palmer and Dunford 2002; Palmer et al. 2001; Stace and Dunphy 2001). The value of dualities thinking lies in the realization that, on the one hand, agility can characterize pockets and teams that have been given the freedom to respond creatively to market developments. Importantly, however, the notion of dualities insists that these pockets cannot exist independent of structures, systems, and boundaries that traditional design structures provide. To exist independently from these is to exist as a random community, not as a goal-directed organization. Resolving the explore—exploit problem leads us to the edge of chaos, not complete chaos. To get to the edge, there has to be a reciprocal force preventing full-blown chaos from taking hold. This is provided by the boundaries of traditional organizing forms, which can safely create pockets of activity primed for emergent innovation in response to market pressure—not only shielded from creativity-destroying bureaucracy, but also bolstered by reliable order and efficiency that can allow the indulgence of freedom in the first place.

Second, dualities in organizing forms assume a dynamic interplay between traditional and new forms in response to contingency variables (Adler et al. 1999; O'Reilly and Tushman 2004; Pettigrew and Whittington 2003; Raynor and Bower 2001; Volberda 1998). Dynamism is a must. It is pointless hoping for innovation unless there is sufficient flexibility in the system to stimulate it in the first place. In this way, duality thinking tempers the extremist logic of hard innovation advocates who assume that novelty can only come from novelty. In fact, we know from at least two decades of research that innovation can occur in parts of conventional bureaucracies that have either deliberately or indiscriminately changed the dynamics of their organizing forms even in subtle ways (Brown and Eisenhardt 1998; Smith 2004; O'Reilly and Tushman 2004).

Third, dualities reflect the twin assumptions of improvisation and learning in the way organizing forms are enacted (Clegg et al. 2002; Lewin and Volberda 1999; March 1991; Stark 2001). In this context, improvisation refers to changes to organizing forms as new information arrives and lessons are learned. As in all aspects of management, the implementation of theory is rarely smooth requiring no remedial action. In reality, managers learn and improvise, as new variables demand modifications.

For example, how does a manager know that he or she has achieved an optimal level of both efficiency and innovation? In truth they probably have no other way of knowing other than through trial and error learning—improvisation—always with a careful eye on performance. Ironically, of course, these are traditional forms of management action, eschewed by hard innovation advocates. Duality thinking encourages the mix of both forms of organizing and the principles that drive them, what Fitzgerald and van Eijnatten (2002) referred to as decisions about 'chaordic' (chaos and order) interaction—the mediation between controllability and responsiveness.

Fourth, the concept of dualities encourages an awareness of interconnectivity within an organization (Child and McGrath 2001; Pettigrew et al. 2003). This means that no part of an organization is independent from activities in others; choices in organizing forms have causal ramifications. Duality thinking encourages research and exploration about causes and effects and does not assume that innovation is a mysterious 'black box' that comes about when creativity somehow gets unleashed within a boundaryless program. The duality paradigm is driven by a desire to know more about ambiguous organizational behavior. Sufficient information about a system gives managers the opportunity to stimulate innovation and preserve efficiency by deliberately creating the right conditions for both at the same time (Smith and Humphries 2004).

Fifth, duality thinking can be applied across all activities an organization undertakes (Schein 1988; Wang and Ahmed 2003; Whittington et al. 1999). For example, dualities can be found in organization structures, processes, and boundaries (Pettigrew et al. 2003). In other words, both efficiency and innovation should be found in all parts of an organization and not just in special programs or as a result of once off initiatives. In the next section, we outline the forthcoming chapter content, explaining how we intend to scrutinize dualities and ambidexterity capability.

LOOKING AHEAD

This book is structured around eight chapters. The following two chapters —two and three—constituting part one of the book, chart out the theoretical basis for dualities in the context of the explore—exploit dilemma and the pursuit of both efficiency and innovation. Part two—Chaps. 4 and 5 —presents the practical realities of creating a dual organization. By using detailed, longitudinal case studies, exposing all the messy problems as well as

all the surprising breakthroughs, we explore how the theory actually translates into practice. Part three, the final two chapters—six and seven—consolidate the lessons emerging from the intersection of theory and practice.
They give form and direction to the implementation process including
another case as well as some key recommendations emanating from our
analyses. Chapter 8 summarizes the three parts, bringing together theory,
practice, and implementation. We will now briefly foreshadow the contents
of each of the remaining seven chapters.

Chapter 2, 'Changing Forms of Organizing,' charts the evolution of
organizing forms giving attention to the nature and scope of exploration
and exploitation as complementary forces stimulating change and continuity, respectively. The chapter aims to reveal the context within which
ambidexterity emerged as a pivotal organizational capability. We define
ambidexterity as an organization's ability to wield existing knowledge while
also creating new knowledge. In a practical sense, an organization possessing ambidexterity capability makes a good return on what it has always
done well while ensuring that it can do new things well in the future. We
propose that ambidexterity capability, untethered and mobilized through
what we describe as a dual organizing forms architecture (explored comprehensively in Chap. 3), provides a mechanism for organizations to
explore and exploit with equal success.

Chapter 3, 'Duality Theory,' ventures further into duality theory as an
overarching but nascent conceptual framework depicting exploration
(change) and exploitation (stability), as neither mutually exclusive nor
inclusive as the long-standing conundrum implies. It begins with a comprehensive review of duality theory and its evolution. Drawing on this
review and ensuing critique of duality characteristics, we argue that ambidexterity capability, underpinned by the five duality characteristics,
reinforces an organization's maintenance of an organizing tension that
delivers both explore and exploit outcomes. Specifically, we propose duality
theory as an explanatory framework and outline five indicative developmental measures for enhancing ambidexterity capabilities.

Chapter 4, 'Embracing the Tension,' furthers the book's proposition
that at the heart of duality theory lays the explore—exploit problem, which
is concerned with how firms can stimulate innovation for the future while
maintaining a high return upon existing opportunities. Based on the longitudinal case study data examined in the chapter, one option involves
pursuing a dual organizational identity embracing innovation and efficiency
as mutually inclusive pursuits. In order to establish what appear to be

contradictory goals, the case firm's leadership successfully implemented an identity transformation encouraging its constituents toward dual explore—exploit capabilities. The chapter's case maps a change process dedicated to successfully marrying the predictable commercial returns of an existing but declining business, and the uncertain pursuit of nascent but high-yield new business. However, what makes this case unique were the tactics employed by executive leaders as they sought to introduce a more entrepreneurial, 'innovation' mindset alongside the traditional business model. The chapter exposes these tactics by applying a sensegiving and sensemaking analysis to the communication between the firm's executive leadership and its partners and directors and, in so doing, shows how they successfully managed an identity transition. The case study therefore reveals how one firm successfully introduced ambidexterity capability by using a method of communication consistent with, and made transparent by, sensegiving and sensemaking. Such data represent a key step in extending duality theory from concept into potential application and testable hypotheses.

Chapter 5, 'Structuring Innovation,' begins by acknowledging the scant empirical work—especially from detailed cases—exposing how first, dual organizing forms can stimulate innovation (explore) without sacrificing efficiency (exploit), and second, how dual organizing forms can deliver a legacy after discontinuation. A second chapter dedicated to case interpretation reveals a complex network of concomitant explore and exploit activities. It depicts a response to the explore—exploit paradox where switching emphasis and resources between the two priorities failed, leading to a novel combination of heavy exploitation-driven actions alongside deep exploration projects. Most pointedly, the organizing response was fluid. This case suggests that part of the case firm's success lay with their fluid organizing forms approach to dealing with the explore—exploit tension. Instead of seeking to delimit it, they sought its escalation into a productive tension, sufficiently powerful to impel individuals to innovate, but sufficiently contained to be captured at an organizational level. According to the literature, a core challenge in reconciling the tension inherent in managing innovation and commercialization structures pivots around a misalignment between the organization and the individual; there has been an overreliance on developing the organizational mechanisms needed to enable dualities with little appreciation for the importance of individual fluidity. Individuals tend to find it difficult to excel at both exploitation and exploration. As a result, they must manage contradictions and conflicting goals, work with uncertainty and ambiguity, be comfortable taking risks, and perform diverse, swiftly changing roles. This chapter highlights a series of organizing forms in which individuals can be innovative and stable, or in other words, develop ambidexterity capability.

Chapter 6, 'Breaking Out,' presents a third case in the series, focusing on an innovation program that 'spun' out, and then back into a parent firm once successful in order to develop scale. The leadership of the firm at the center of this study made a bold commitment to making innovation a core feature of their culture. At the same time, they recognized the imperative to maintain a controlled, commercial yield from the firm's mature and well-respected services. Between this 'explore—exploit' tension, the firm developed an innovation program designed to stimulate new, 'ahead of the curve,' services in order to secure a steady turnover of fresh revenue. The central initiative involved the creation of a fast-tracked, technology-focused, product-oriented business unit that turned the traditional services firm business model upside-down by using low touch services, remote connections, self-service clients, and a high volume of transactions and clients, all within low margins. This chapter charts the life cycle of the new spin-out. From a practical perspective, this case suggests that one way to take advantage of the explore—exploit tension involves the use of heterogeneous, fluid groups positioned to leverage a co-design model centering around rapid prototyping undertaken in collaboration with 'lighthouse' pilot clients. The case illustrates how a firm can span the troublesome implementation gap between theoretical notions of explore—exploit and the successful delivery of innovation and commercial efficiency at the same time.

Chapter 7, 'Designing Innovation,' considers how new thinking and tools can augment ambidexterity capability. Over the past decade or so, design thinking has received increasing attention both within and beyond contemporary design discourse. In particular, management and business education has feted design thinking as both important and revolutionary in how it can change the way we perceive, articulate, and solve complex problems that tend to be considered beyond rational pragmatism. In this chapter, we discuss observations that design's role in business innovation has been narrow and limited to creative and technical domains. We argue that a reductionist view of design fails, occurring when managers confuse design thinking with the use of design tools. In addition, we establish some concrete recommendations for developing ambidexterity capability based on a combination of our case lessons, results from salient research, and our experience in melding the two. We conclude that ambidexterity capability accompanies a human-centered, fuzzy, and fluid approach to organizing in general, and innovation in particular.

The final chapter (Chap. 8), 'The Efficient Innovator,' provides some concluding observations about how to take advantage of the explore—exploit tension. It comments on the use of heterogeneous, agile communities capable of working in the gray area, where the normal rules and expectations become more elastic, and novel propositions can be tested in the real world quickly at a low cost and risk. Further, a firm must shift from 'and/ or' approaches to reconciling exploit—explore tensions, instead optimizing both by explicitly affording everyone the right to innovate, while implicitly fostering everyone's responsibility to innovate. Vital to success is the commitment of an organization's leadership group to drive such a mindset. The new mindset signals a commitment to rapid prototyping and concept proofs in the market combined with a design-oriented, user-based mode of thinking about client experiences. At the same time, lengthy commercialization plans and off-the-shelf service solutions are discarded. We maintain that an innovation ecosystem shows how structure and formalization can augment the conversion of ideas to commercial implementation, without compromising speed. Through such methods, expert and developing innovators fuel incremental innovation while consolidating innovation into a firm's conventional business units. As a result, the chapter finishes by once again calling for the reinvention of innovation thinking, mobilized through ambidexterity capabilities.

Conclusion

The arguments for new forms of organizing, or those embracing more flexibility and agility, are based on the view that environmental uncertainty is incompatible with traditional or bureaucratic organizational forms. Accordingly, network-driven, flat, permeable forms of organization hold significant advantages over hierarchical, rule-centered bureaucracies. Although intuitive, this approach did not deliver more innovation, or at least it did so at the expense of maximizing current returns. And so, as we observed at the outset of this chapter, an organizing form dilemma has emerged. On the one hand, traditional forms of organizing are efficient but inflexible, while on the other hand, new forms of organizing are adaptive but disorderly.

Reinventing innovation means thinking differently about the relationship between doing old things and doing new things in organizations.

Instead of trading off what appear to be opposites, in this book we argue for the power of stability *with* change. We suggest that innovation does not necessarily mean focusing only on exploring the new. Rather, we reinvent innovation to mean an imaginative transformation of the way organizations conceive the balance between exploration and exploitation. We discard with the conventional view of one or the other, as well as an equitable balance between the two. What we need is lots of both—a proficiency we summarize as ambidexterity capability.

REFERENCES

Adler, P. S., Goldoftas, B., & Levine, D. I. (1999). Flexibility versus efficiency? A case study of model changeovers in the Toyota production system. *Organization Science, 10*(1), 43–68.

Bahrami, H. (1992). The emerging flexible organization: Perspectives from Silicon Valley. *California Management Review, 34*(4), 33–52.

Brown, S. L., & Eisenhardt, K. M. (1998). *Competing on the edge: Strategy as structured chaos.* Boston, MA: Harvard Business School Press.

Child, J., & Mcgrath, R. (2001). Organizations unfettered: Organizational form in an information-intensive economy. *The Academy of Management Journal, 44*(6), 1135–1148.

Clegg, S., da Cunha, J. V., & e Cunha, M. P. (2002). Management paradoxes: A relational view. *Human Relations, 55*(3), 483–503.

Dijksterhuis, M. S., Van Den Bosch, F., & Volberda, H. W. (1999). Where do new organizational forms come from? Management logics as a source of coevolution. *Organization Science, 10*(5), 569–582.

Dimaggio, P. (Ed.). (2001). *The twenty-first century firm.* Princeton, NJ: Princeton University Press.

Dunphy, D., & Stace, D. (1993). *Under new management: Australian organisations in transition.* Sydney: McGraw-Hill.

Eisenhardt, K. M., & Brown, S. L. (1998). Competing on the edge: Strategy as structured chaos. *Long Range Planning, 31*(5), 786–789.

Evans, P. (1999). HRM on the edge: A duality perspective. *Organization, 6*(2), 325–338.

Fenton, E., & Pettigrew, A. (2000). *Theoretical perspectives on innovative forms of organizing.* London: Sage.

Fenton, E. M., & Pettigrew, A. M. (2001). Theoretical perspectives on new forms of organizing. In A. M. Pettigrew & E. M. Fenton (Eds.), *The innovating organization* (pp. 1–46). London: Sage.

Fitzgerald, L., & Van Eijnatten, F. (2002). Reflections: Chaos in organizational change. *Journal of Organizational Change Management, 15*(4), 402–411.

Ghoshal, S., & Bartlett, C. A. (1995). Building the entrepreneurial corporation: New organizational processes, new managerial tasks. *European Management Journal, 13*(2), 139–155.

Giddens, A. (1984). *The constitution of society: Outline of the theory of structuration.* California: University of California Press.

Graetz, F., & Smith, A. (2009). Changing forms of organizing in Australian public companies. *Asia Pacific Journal of Human Resources, 47*(3), 340–360.

Jackson, W. (1999). Dualism, duality and the complexity of economic institutions. *International Journal of Social Economics, 26*(4), 545–558.

Jonsson, S. (2000). Innovation in the networked firm: The need to develop new types of interface competence. In J. Birkinshaw & P. Hagström (Eds.), *The flexible firm: Capability management in network organizations* (pp. 106–125). New York: Oxford University Press.

Kelly, K. (1998). *New rules for the new economy.* New York: Viking.

Lewin, A., & Volberda, H. (1999). Prolegomena on coevolution: A framework for research on strategy and new organizational forms. *Organization Science, 10*(5), 519–534.

Lewin, A. Y., Long, C., & Carroll, T. (1999). The coevolution of new organizational forms. *Organization Science, 10*(5), 535–550.

Limerick, D., & Cunnington, B. (1993). *Managing the new organization.* Sydney: Business and Professional Publishing.

March, J. G. (1991). Exploration and exploitation in organizational learning. *Organization Science, 2*(1), 71–87.

Nohria, N., & Berkley, J. D. (1994). The virtual organization: Bureaucracy, technology, and the implosion of control. In C. Heckscher & A. Donnellon (Eds.), *The post-bureaucratic organization: New perspectives on organizational change* (pp. 108–128). Thousand Oaks, CA: Sage.

O'Reilly, C. A., & Tushman, M. L. (2004). The ambidextrous organization. *Harvard Business Review, 82*(4), 74–81.

Palmer, I., & Dunford, R. (2002). Out with the old and in with the new? The relationship between traditional and new organizational practices. *International Journal of Organizational Analysis, 10*(3), 209–226.

Palmer, I., & Hardy, C. (2000). *Thinking about management: Implications of organizational debates for practice.* London: Sage.

Palmer, I., Dunford, R., Rura-Polley, T., & Baker, E. (2001). Changing forms of organizing: Dualities in using remote collaboration technologies in film production. *Journal of Organizational Change Management, 14*(2), 190–212.

Pettigrew, A., & Whittington, R. (2003). Complementarities in action: Organizational change and performance in BP and Unilever 1985–2002. In A. Pettigrew, R. Whittington, L. Melin, C. Sanchez-Runde, F. Van Den Bosch, W. Ruigrok, & T. Numagami (Eds.), *Innovative forms of organizing* (pp. 173–207). London: Sage.

Pettigrew, A. M., Whittington, R., Melin, L., Sanchez-Runde, C., Van Den Bosch, F., Ruigrok, W., et al. (2003). *Innovative forms of organizing*. London: Sage.

Pettigrew, A. M. & Fenton, E. M. (Eds.). (2000). *The innovating organization*. London: Sage.

Quinn, J. B., Anderson, P., & Finkelstein, S. (1998). New forms of organizing. In H. Mintzberg & J. Quinn (Eds.), *Readings in the strategy process* (2nd ed., pp. 162–174). Upper Saddle River, NJ: Prentice Hall.

Raynor, M. E., & Bower, J. L. (2001). Lead from the center: How to manage divisions dynamically. *Harvard Business Review, 79*(5), 93–100.

Rindova, V. P., & Kotha, S. (2001). Continuous 'morphing': Competing through dynamic capabilities, form and function. *The Academy of Management Journal, 44*(6), 1263–1280.

Sanchez-Runde, C. J., & Pettigrew, A. M. (2003). Managing dualities. In A. M. Pettigrew, R. Whittington, L. Melin, C. Sanchez-Runde, F. Van Den Bosch, W. Ruigrok, & T. Numagami (Eds.), *Innovative forms of organizing* (pp. 243–250). London: Sage.

Schein, E. (1988). *Organizational psychology* (3rd ed.). Englewood Cliffs, NJ: Prentice Hall.

Schilling, M. A., & Steensma, H. K. (2001). The use of modular organizational forms: An industry-level analysis. *The Academy of Management Journal, 44*(6), 1149–1168.

Smith, A. (2004). Complexity theory and change management in sport organizations. *Emergence: Complexity and Organization, 6*(1–2), 70–79.

Smith, A., & Graetz, F. (2006). Organizing dualities and strategizing for change. *Strategic Change, 15*(5), 231–239.

Smith, A., & Humphries, C. (2004). Complexity theory as a practical management tool: A critical evaluation. *Organization Management Journal, 1*(2), 91–106.

Stace, D., & Dunphy, D. (2001). *Beyond the boundaries: Leading and recreating the successful enterprise* (2nd ed.). Sydney: Mcgraw-Hill.

Stark, D. (2001). Ambiguous assets for uncertain environments: Heterarchy in postsocialist firms. In P. Dimaggio (Ed.), *The twenty-first century firm* (pp. 69–104). Princeton, NJ: Princeton University Press.

Tetenbaum, T. (1998). Shifting paradigms: From Newton to chaos. *Organizational Dynamics, 26*(4), 21–32.

Ulrich, D., & Wiersema, M. F. (1989). Gaining strategic and organizational capability in a turbulent business environment. *The Academy of Management Executive, 3*(2), 115–122.

Volberda, H. W. (1998). *Building the flexible firm*. Oxford: Oxford University Press.

Wang, C., & Ahmed, P. (2003). Structure and structural dimensions for knowledge-based organizations. *Measuring Business Excellence, 7*(1), 51–62.

Whittington, R., Pettigrew, A., Peck, S., Fenton, E., & Conyon, M. (1999). Change and complementarities in the new competitive landscape: A European panel study, 1992–1996. *Organization Science, 10*(5), 583–600.

CHAPTER 2

Changing Forms of Organizing

Abstract This chapter charts the evolution of organizing forms, in particular the nature and scope of exploration and exploitation as complementary forces stimulating change and continuity. The chapter contextualizes the emergence of ambidexterity as the capacity to both use and refine existing knowledge while also creating new knowledge. Further, ambidexterity capability may be leveraged through a duality-based organizing forms architecture, providing the means for enabling organizations to explore and exploit with equal success. Finally, the chapter highlights how ambidexterity capability introduces a unique process view of the explore—exploit tension that draws on duality theory as an explanatory framework.

Keywords Organizing forms · Ambidexterity capability

INTRODUCTION

Chapter 1 introduced the commentary beginning in the 1950s calling for 'new' forms of organizing. In the context of a volatile environment escalating in complexity and turbulence, a new popular message radiated through the late twentieth century. The need for an overhaul of traditional, functionally specialized, hierarchical systems and structures had arrived. No doubt traditional forms had served organizations admirably during less demanding and more certain times. However, a dynamic and ambiguous business environment sat uncomfortably with the newfound need for continuity and stability. Growth and prosperity became connected to a

© The Author(s) 2017
A.C.T. Smith et al., *Reinventing Innovation*,
DOI 10.1007/978-3-319-57213-0_2

flexible, responsive approach, primed to expect the unexpected (Beer and Walton 1990). Organizations were therefore urged to overthrow rigid systems and structures in favor of flat, loosely assembled, networked forms of organizing. As the cumbersome old models languished, the new were hailed as the magic bullet for innovation, exploration, and learning (Limerick and Cunnington 1993; Kanter et al. 1992; Naisbitt and Aburdene 1990; Ulrich and Wiersema 1989).

Claims for more flexible, agile forms of organizing presuppose that environmental turbulence requires a certain structural response that traditional organizational forms, built around steady-state conformity, are ill-equipped to address. But, despite all the noise about discarding tight traditional structures for fluid new ones, organizations actually shifted differently. Studies on organizations across the Western world, for example, have revealed that while new, more flexible forms of organizing did emerge within organizations, hierarchy and other traditional organizational practices remained steadfastly in place. As it turned out, most organizations discovered that a direct exchange of old for new was not the innovation panacea that was promised. In fact, the core aspects of traditional structure—planning, coordinating, and direction-setting—also provide stability during periods of uncertainty. As recent studies show (and we demonstrate in detail through our case study analyses in Chaps. 4–6), the *exploitative* 'accountability and control' attributes of traditional forms of organizing complement and support the *explorative*, 'adaptive attributes' of new forms of organizing (O'Reilly and Tushman 2004; Palmer and Dunford 2002; Pettigrew et al. 2003; Raynor and Bower 2001; Volberda 1998). The problem seems to lie with how we look at the two kinds of organizing frameworks.

Rather than viewing the dual imperatives of change and continuity as distinctive but complementary elements of organizational design, they are generally treated as mutually exclusive, antagonistic opposites. As such they are often represented in adversarial continua such as control versus freedom, hierarchy versus empowerment, or stability versus flexibility. However, by favoring only one side of the change-continuity equation, organizations risk either paralysis through inaction, or chaos through overreaction (Weick 1979). As Davis and Lawrence (1977) noted, the dilemma of change comes with the dilemma of an 'either-or world.'

Some commentators view the 'either-or' dilemma as a problem of managing 'organizing tensions' (Asch and Salaman 2002). Some organizing tension discussions focus on pressures to differentiate *and* integrate, as well as the need for controllability *and* responsiveness. Others argue that the way forward involves constructing the so-called ambidextrous forms encapsulating explore and exploit domains because they provide an ideal architecture for simultaneous tight and loose structures (Benner and Tushman 2003). Although it sounds awkward, a tight and loose arrangement means that flexible structures operate inside, beside, or outside hierarchical bureaucracy. Yet both kinds of organizing seem essential because flexible exploration goes off the radar without some exploitation to control and scale its yield. Conversely, commercial exploitation inevitably runs out unless a pipeline of innovative products is developed.

According to Smith and Tushman (2005, p. 523), exploitation can be seen as 'variance decreasing' based on 'disciplined problem solving.' Exploitation is concerned with continuity and control—the 'tight' side of the organizing equation. It draws on and builds from an organization's past, aiming to increase efficiency and profitability in the current business model. Exploration, in contrast, constitutes 'variance increasing' through trial and error experimentation. Exploration depicts the 'loose,' responsive organizing mode. It is concerned with change and adaptability—encouraging creativity and risk-taking, and tapping into previously untested markets and opportunities (Groysberg and Lee 2009; Smith and Tushman 2005).

Discussion around these concomitant, yet competing, dual pressures gave rise to the concept of ambidextrous forms of organizing. Ambidextrous forms of organizing were seen to reflect a 'mix and match' approach (Jackson and Harris 2003), comprising interactions between organic-, mechanistic-, and knowledge-based structures (Wang and Ahmed 2003). Organizational ambidexterity is really just the capacity to simultaneously exploit existing capabilities while exploring new opportunities (Tushman and O'Reilly 1996). It sounds straightforward enough—even a natural state for a healthy organization—but has proven to be an elusive combination.

This chapter charts the evolution of organizing forms, in particular the nature and scope of exploration and exploitation as complementary forces stimulating change and continuity. We seek to contextualize the emergence of ambidexterity as an organizational capability. As Turner et al. (2013, p. 320) explained, ambidexterity represents the capacity 'to both use and refine existing knowledge (exploitation) while also creating new knowledge to overcome knowledge deficiencies or absences identified with the execution of the work (exploration).' We propose that *ambidexterity capability*, leveraged through what we describe as a dual organizing forms architecture (explored comprehensively in Chap. 3), may provide the means for enabling organizations to explore and exploit with equal success. This provides the background for the ensuing discussion in Chap. 3. It highlights how ambidexterity capability introduces a unique process view of the explore—exploit tension that draws on duality theory as an explanatory framework. But first a little history.

THE EVOLUTION OF ORGANIZING FORMS: TOWARD EXPLORATION AND EXPLOITATION

Perceptions had shifted inexorably by the turn of the century. As concepts like sustainable futures, long-term growth, competitive collaboration, and people-focused management came into vogue as part of the organizing lexicon through the last decades of the twentieth century, the call for a response encompassing both continuity and change gathered momentum (Limerick and Cunnington 1993; Kanter et al. 1992; Naisbitt and Aburdene 1990; Ulrich and Wiersema 1989). Where earlier environmental turbulence appeared infrequent and manageable for organizations with sufficient resolve, the new century quickly became synonymous with constant change. At the same time, innovation remained slippery. Even its successful delivery could not save firms that had not maintained high levels of core business efficiency as well.

Organizations were caught in a vicious circle. The formulaic model of traditional structures and systems, which focused on planning and budgeting, and putting in place clearly defined systems and procedures (Evans and Doz 1992), had become cumbersome. They simply did not work well in a complex, highly competitive, and globalized marketplace. As Pettigrew and Fenton (2000, p. 279) explained, the overriding assumption was that the 'new competitive landscape renders anachronistic traditional, efficiency-oriented vertical structures and triggers the search for new organizational practices in which flexibility, knowledge collecting and connecting, and horizontal collaboration are essential characteristics.' A seismic shift in focus was needed from a solitary concentration on stability and control to managing innovation and change as well

As we have noted, however, despite some enthusiastic predictions, this shift did not spell the demise of bureaucracy. Rather, high-performing organizations adopted dual organizing forms drawing on the strengths of traditional, robust forms of organizing while they experimented with the new (O'Reilly and Tushman 2004; Pettigrew et al. 2003; Palmer et al. 2007;

Palmer and Dunford 2002; Raynor and Bower 2001; Volberda 1998). The resulting 'dual' or 'paired' forms blended familiar formalized, traditional work practices with new, agile and flexible work arrangements. Instead of just trying to ride out uncertain times and dicey markets, firms tried a blended model in order to maximize control and flexibility at the same time (O'Reilly and Tushman 2004; Pettigrew et al. 2003). Paradoxically, some bureaucratic rules proved essential in facilitating organizational agility and adaptability. Rigid structures offer a stable platform from which exploration and experimentation can be safely and expeditiously pursued (Bahrami 1992; Brooks and Saltzman 2016). It was this realization—stability and change can be combined in high measures to produce an optimal outcome —that opened the door to an entirely new way of thinking about meeting explore and exploit targets without compromising one or the other.

Powered by a new paradigm, the marriage of traditional with new forms of organizing fed into a series of studies investigating dualities (Evans and Doz 1989; Quinn and Cameron 1988), organizational ambidexterity (Gibson and Birkinshaw 2004; O'Reilly and Tushman 2004; Tushman and O'Reilly 1996), and the explore—exploit tension (March 1991). While these organizing/structuring concepts only gained prominence in the last decades of the twentieth century with the pioneering work of researchers like Tushman and O'Reilly, Evans and Doz, and March, their provenance can be traced back to the early 1960s. In fact, scholars such as Burns and Stalker (1961), Lawrence and Lorsch (1967), and Duncan (1976) first coined the term 'the ambidextrous organization.' We think that as a broader, more inclusive concept, 'dualities' or 'duality theory,' perhaps offers the most productive foundational framework.

Duality theory presents an overarching but nascent conceptual framework depicting exploration (change) and exploitation (stability) as neither mutually exclusive nor inclusive, as the long-standing conundrum implies. Rather, the two exist as a 'duality,' a non-reductive construct characterized by simultaneity, relatedness, minimal thresholds, dynamism, and improvisation (Evans and Doz 1992; Graetz and Smith 2008; Hedberg et al. 1976). In other words, an organizing duality exists when a unique combination of high levels of both change and stability occurs simultaneously, taking on an emergent, complex dynamic. Like a champion sporting team, an organization with duality characteristics can both attack aggressively and defend indefatigably without shifting players or discussing tactics. Thus, we talk about such organizations as possessing ambidextrous capabilities, pressing forward to score as easily as holding resolute in defense. They make good now while working on what will be good later.

Organizing forms and innovation literature promote the merits of ambidextrous forms of organizing, the inescapable need for organizational dualities, and the possibilities of pursuing concurrent exploration and exploitation. Faced with such an array of terms all expounding a 'paired' or double-pronged approach, the challenge for scholars and practitioners became one of meanings. The salient issue revolved around whether ambidexterity, dualities, and explore—exploit all referred to the same thing, or whether they should have been viewed as either interrelated or distinct. Here, we argue that the successful pursuit of both exploration and exploitation relies first on nurturing ambidexterity capability. To that end, we view ambidexterity capability as an essential pre-condition for explore—exploit outcomes because it normalizes the seemingly contradictory, divisive tension that accompanies the pursuit of exploration and exploitation.

By 'normalization' we mean that ambidexterity capability enables organizations not only to recognize that the tension is real, but also to learn to live with and accept contradiction. In essence, contradiction (or the presence of complexity) is a 'normal', albeit critical, component of a healthy organization. Complementarities emerge through engagement with coexisting (seemingly contradictory) forces, such as stability and change. While their association is undeniable, they are not substitutable. Similarly, dualities, as the word suggests, refers to the distinctive (but complementary) organizing modes present in the explore—exploit duality, such as loose/tight coupling, empowerment/leadership, action/planning, and flexibility/control, and are summarized in Table 2.1. Notions of ambidexterity, dualities, and explore—exploit were all invented in order to best search for a mode of structuring and organizing to foster innovation, flexibility, experimentation, and organizational learning without compromising efficiency.

ORGANIZATIONAL AMBIDEXTERITY: BUILDING THE CAPACITY TO EXPLORE AND EXPLOIT

For researchers investigating organizational ambidexterity, and practitioners seeking to cultivate it, the key question concerns how to manage the ostensibly contradictory elements of continuity (efficiency) and change (flexibility) with equal dexterity (Tushman et al. 2010). Organizational ambidexterity underpins survival in dynamic, complex environments, 'aligned and efficient in their management of today's business demands,

Table 2.1 Characteristics of exploration and exploitation

Exploit domain *Exploitative continuity goals* *Control and efficiency*	*Explore domain* *Explorative change goals* *Innovation and flexibility*
Individual focus • Focus on individual skills, expertise, personal characteristics, specialization, competition	Organizational focus • Focus on cooperation and collaboration through teamwork and cross-functional work groups • Building shared values, trust
Static • Maintain the status quo • Assumes a stable, predictable environment • Use of traditional KPIs: balance sheet, gross margins, profit and loss, salary cost, etc.	Dynamic • Organic, flexible, adaptive, loosely coupled • Foster risk-taking and creativity • Use of distinctive benchmarks and performance measures to encourage innovation
Internal focus • Inward, closed system focus • Ownership of upstream and downstream resources and capabilities; insourcing • Long-term staffing—job for life	External focus • Open systems focus, monitor external environment • Short-term staffing: core of permanent staff; buy in expertise for short term projects • Outsource 'non-core' activities and services
Differentiation • Functional specialization, silos—focus on specific, individual skills • Aggregation of business units—traditional divisional, semi-autonomous business structures	Integration • Collaborative networks: Strategic partnerships, joint ventures to complement in-house capabilities and resources • Disaggregation of business units: emphasizing both autonomy and interdependence
Hierarchy • Tall pyramid structures • Tightly coupled, top-down control • Low discretionary powers at operational level	Teamwork • Cross-functional project-based work groups/small specialized teams—autonomy and interdependence • Removing middle management/supervisory layers, increasing individual autonomy and discretion
Leadership/command • Strong, top-down leadership and supervision. Directive	Empowerment • Distributed leadership. Delegation of decision making, increased self-direction and self-management

(continued)

Table 2.1 (continued)

Exploit domain Exploitative continuity goals Control and efficiency	Explore domain Explorative change goals Innovation and flexibility
leadership. Authority and power with senior management	
Accountability, Control	Freedom, Flexibility
• Focus on uniformity, standardization, performance measurement and quality control systems to monitor, coordinate, control, and disseminate information	• Recognition that increased competition at both the local and global levels calls for increased responsiveness and flexibility, and continuous improvement and innovation
Planning	Action
• Traditional, rational, systematic approach to strategic planning. Assumes a stable, predictable environment	• Strategy as emergent, unfolding, responding proactively to environmental uncertainty. Focus on doing
Centralization	Decentralization
• Centralized systems and decision making: authority; power with senior management	• Redistribution of decision making to involve employee or worker participation
Tight Coupling	Loose coupling
• Allied to accountability and control. Rigid structures, with traditional segmentation of skills and knowledge into discrete departments. Functional silos	• Fluid, permeable boundaries simultaneously asserting 'both autonomous distinctiveness and interdependence,' thus enabling differentiation and integration between business units

while also adaptive enough to changes in the environment that they will still be around tomorrow' (Gibson and Birkinshaw 2004, p. 209). In short, the capacity to manage contradictory tensions simultaneously—such as efficiency and innovation—means that an organization can extract an efficient return on its long-term investments while chasing new and risky innovations along the way (Andriopoulos and Lewis 2009; Nosella et al. 2012; Turner et al. 2013). Efficiency ensures short-term prosperity, mitigating the risks and expenses of innovation. Conversely, a well-stoked innovation pipeline secures long-term viability and profitability.

Central to ambidexterity is the capacity for 'simultaneity' in managing organizing tensions (Andriopoulos and Lewis 2009; Lubatkin et al. 2006;

Turner et al. 2013). Four tensions in particular take precedence (Raisch et al. 2009): differentiation/integration; individual/organizational; static/dynamic; and internal/external. These tensions represent related and complementary rather than opposing mechanisms for organizational performance. For example, instead of a balanced compromise, efficiency and flexibility should be each maximized. While exploration and exploitation demand distinct learning processes, success arrives from simultaneity in execution (March 1991). Although March (1991) did not employ the term ambidexterity to describe dual organizing forms, his contribution led to an acceptance that both sides must be considered as a complex and dynamic unit (Turner et al. 2013). Ambidexterity is about doing both exploration and exploitation, thereby enabling an organization to 'reconfigure existing assets and capabilities to sense and seize new opportunities' (O'Reilly and Tushman 2008, p. 200).

Ambidexterity capability overcomes the need for trade-offs in allocating scarce resources to either exploitation or exploration at the expense of the other. We propose that *building* ambidexterity capability drives the uncompromising pursuit of exploration and exploitation. As Hill and Birkinshaw (2014, p. 1900) concluded in their study of corporate venture (CV) units, it was the capacity to 'reconcile the competing demands for exploration and exploitation' that proved critical to survival.

EXPLORATION AND EXPLOITATION: EMBRACING THE DUALISTIC TENSION

In his seminal paper on exploration and exploitation in organizational learning, March (1991, p. 85) defined the essence of exploitation as 'the refinement and extension of existing competences, technologies, and paradigms' with 'positive, proximate, and predictable' returns. In contrast, he described exploration as 'experimentation with new alternatives with returns that are 'uncertain, distant, and often negative.' The ever-present challenge for organizations remains how 'to engage in sufficient exploitation to ensure its current viability and, at the same time, devote enough energy to exploration to ensure its future viability' (Levinthal and March 1993, p. 105). Doing well now and doing well later have never been considered compatible objectives. Compounding this dilemma, many senior managers harbor an instinctive discomfort with ambiguity and uncertainty, encouraging an intuitive reluctance to adapt and improvise. Most naturally seek order and predictability through formulaic procedures

(Collins 2003). Yet, as Pascale (1990) suggested, it is the tension or 'dynamic synthesis' between contradictory opposites that provides the catalyst for self-renewal and long-term organizational effectiveness.

Traditional, formalized logic based on either/or thinking would try and *resolve* the dualistic tension between what were once viewed as 'conflicting truths' (Lewis 2000, p. 761) by favoring one option at the expense of the other. In contrast, we maintain that developing ambidexterity capabilities allows organizations to convert the tensions into explore—exploit outcomes. Yet, although many leaders accept that exploration and exploitation demand equal, uncompromising pursuit, an assumption still holds that each goal takes an organization in a different direction.

We propose that leaders and managers should think in terms of developing ambidexterity capabilities. Here, action reflects the merger of the two explore—exploit polarities. Instead of saying what a company needs to *do* (explore—exploit), the ambidexterity construct focuses on what capabilities a company needs to become comfortable with in pursuing both exploration and exploitation. While this chapter exclusively focuses on the theoretical background to explore—exploit in the form of dualities, our later chapters provide detailed cases unpacking how ambidexterity capabilities can be achieved.

Dual Explore—Exploit Forms of Organizing: The Ambidexterity Capability Pre-requisite

As organizations engage in more cross-boundary collaborations with both internal and external stakeholders, dualities highlight the need to manage 'loose-tight' relationships. Organizations need to establish structures that enhance flexibility and responsiveness (exploration) while simultaneously bolstering performance efficiencies (exploitation). Lowendahl and Revang (2004, p. 50) echoed this sentiment advocating that 'modern change processes demand simultaneous management of change in some dimensions and stability in others' in order to realize the dual need for efficiency and flexibility. As we have noted, inconsistent yet interrelated organizing forms (Lewis 2000) (such as flexibility-efficiency, accountability-freedom, control-innovation, and dynamic-static) have been linked to exploration and exploitation (He and Wong 2004; Lewis 2000; March 1991; O'Reilly and Tushman 2008). Tushman and O'Reilly (1996) defined the capacity to simultaneously exploit existing capabilities while exploring new opportunities as a theory of 'organizational ambidexterity.'

Despite, or perhaps because of, this definition, the labels 'exploration and exploitation' and 'organizational ambidexterity' have appeared almost interchangeably in the literature. It reflects a colloquial presumption that organizational ambidexterity and exploration-exploitation are at least co-dependent if not interchangeable concepts (Benner and Tushman 2003; Chen and Miller 2010; Groysberg and Lee 2009; He and Wong 2004; Lowendahl and Revang 2004; Tushman et al. 2010). Writing more recently, Voss and Voss (2013) observed that achieving ambidexterity is particularly challenging for smaller, younger organizations that lack the resources, capabilities, or experience to simultaneously pursue exploration and exploitation. The correspondence is significant to our argument that undertaking exploration and exploitation with equal vigor and determination depends upon building ambidexterity capability.

We think that organizational ambidexterity and exploration—exploitation are *not* interchangeable. Rather, ambidexterity capability precedes the successful, simultaneous pursuit of exploration and exploitation. Together they deliver distinctive yet complementary, co-dependent elements of organizing and structuring. As explained in Chap. 3, ambidexterity capability allows organizations to embed organizational dualities within conventional operations as measured by their ability to deliver both explore and exploit outcomes. As Jansen et al. (2008) showed, while descriptive accounts are plentiful, few studies have tried to explain the drivers of ambidexterity, although the most recent are addressing its 'mechanisms' (Andriopoulos and Lewis 2009; Jansen et al. 2012; Lubatkin et al. 2006; Turner et al. 2013). While there is limited research on how to manage ambidexterity or the explore—exploit tension, 'almost all agree that ambidexterity is an organizational capability that makes it possible to resolve tensions that arise in organizations' (Nosella et al. 2012, p. 540). In contrast, we argue that the aim of building ambidexterity capability is *not* to resolve tension but to encourage it in the form of explore—exploit outcomes.

CONCLUSION

As this chapter has reinforced, the exploitation—exploration debate is alive and well. Commentaries on balancing or reconciling the organizing tensions accompanying continuity and change thread through more than 50 years of literature. The enduring argument holds that sustainable long-term organizational futures demand simultaneous exploration and exploitation, pursued with equal vigor. In fact, the simultaneous pursuit of exploration— invoking innovation and efficiency—premised by stability and control remains critical to an organization's success and survival. No one challenges the new realities of organizational life where tension and discontinuity are natural, irreducible elements (Gibson and Birkinshaw 2004; He and Wong 2004; March 1991; O'Reilly and Tushman 2004, 2008; Tushman et al. 2010). The question has always been what to do about it. We have suggested that ambidexterity capability constitutes the essential precondition for explore—exploit simultaneity and its subsequent execution. Instead of saying what a company needs to *do* (explore and exploit), ambidexterity focuses on what capabilities a company needs to *build* in order to grow and develop, bringing together both structural and contextual considerations. Rather than just assuming that explore and exploit outcomes are both needed, attention must shift to the development and sustenance of exploration and exploitation. As this indicates, the aim of building ambidexterity capability is *not* to resolve, but to engage with, the tension between the dual explore—exploit elements. In addition, we have referenced duality theory as a potential explanatory framework for understanding and managing the tensions that arise in building a capacity for simultaneous, dynamic, related exploration and exploitation through the medium of organizational ambidexterity. These lines of inquiry are pursued in Chap. 3.

References

Andriopoulos, C., & Lewis, M. W. (2009). Exploitation-exploration tensions and organizational ambidexterity: Managing paradoxes of innovation. *Organization Science, 20*(4), 696–717.

Asch, D., & Salaman, G. (2002). The challenge of change. *European Business Journal, 14*(3), 133–143.

Bahrami, H. (1992). The emerging flexible organization: Perspectives from Silicon Valley. *California Management Review, 34*(4), 33–52.

Beer, M., & Walton, E. (1990). Developing the competitive organization: Interventions and strategies. *American Psychologist, 45*(2), 154–216.

Benner, M. J., & Tushman, M. L. (2003). Exploitation, exploration, and process management: The productivity dilemma revisited. *Academy of Management Review, 28*(2), 238–256.

Brooks, S. M., & Saltzman, J. M. (2016). *Creating the vital organization*. New York: Palgrave Macmillan.

Burns, T., & Stalker, G. M. (1961). *The management of innovation*. London, UK: Tavistock.

Chen, M.-J., & Miller, D. (2010). West meets East: Toward an ambicultural approach to management. *Academy of Management Perspectives, 24*(1), 17–22.

Collins, D. (2003). Guest editor's introduction: Re-imagining change. *Tamara: Journal of Critical Postmodern Organization Science, 2*(4), iv–xi.

Davis, Stanley M., & Lawrence, P. R. (1977). *Matrix.* Reading, MA: Addison-Wesley.

Duncan, R. (1976). The ambidextrous organization: Designing dual structures for innovation. In R. Kilmann, L. Pondy, & D. Slevin (Eds.), *The management of organization* (pp. 167–188). New York: North-Holland.

Evans, P., & Doz, Y. (1989). The dualistic organization. In P. Evans, Y. Doz, & A. Laurent (Eds.), *Human resource management in international firms: Change, globalization, innovation* (pp. 219–242). London, UK: Macmillan.

Evans, P., & Doz, Y. (1992). Dualities: A paradigm for human resource and organizational development in complex multinationals. In V. Pucik, N. Tichy, & C. Barnett (Eds.), *Globalizing management: Creating and leading the competitive organization* (pp. 85–106). New York: Wiley.

Gibson, C. B., & Birkinshaw, J. (2004). The antecedents, consequences, and mediating role of organizational ambidexterity. *Academy of Management Journal, 47*(2), 209–226.

Graetz, F., & Smith, A. (2008). The role of dualities in arbitrating continuity and change in forms of organizing. *International Journal of Management Reviews, 10*(3), 265–280.

Groysberg, B., & Lee, L.-E. (2009). Hiring stars and their colleagues: Exploration and exploitation in professional service firms. *Organization Science, 20*(4), 740–758.

He, Z. L., & Wong, P. K. (2004). Exploration vs. exploitation: An empirical test of the ambidexterity hypothesis. *Organization Science, 15*(4), 481–494.

Hedberg, B., Nystrom, P., & Starbuck, W. H. (1976). Camping on seesaws: Prescriptions for a self designing organization. *Administrative Science Quarterly, 21*(1), 41–65.

Hill, S. A., & Birkinshaw, J. (2014). Ambidexterity and survival in corporate venture units. *Journal of Management, 40*(7), 1899–1931.

Jackson, P., & Harris, L. (2003). E-business and organizational change: Reconciling traditional values with business transformation. *Journal of Organizational Change Management, 16*(5), 497–511.

Jansen, J. J. P., George, G., Van den Bosch, F. A. J., & Volberda, H. W. (2008). Senior team attributes and organizational ambidexterity: The moderating role of transformational leadership. *Journal of Management Studies, 45*(5), 982–1007.

Jansen, J. J. P., Simsek, Z., & Cao, Q. (2012). Ambidexterity and performance in multiunit contexts: Cross-level moderating effects of structural and resource attributes. *Strategic Management Journal, 33*(11), 1286–1303.

Kanter, R. M., Stein, B. A., & Jick, T. D. (1992). *The challenge of organizational change.* New York: The Free Press.

Lawrence, P. R., & Lorsch, J. W. (1967). Differentiation and integration in complex organizations. *Administrative Science Quarterly, 12*(1), 1–47.

Levinthal, D. A., & March, J. G. (1993). The myopia of learning. *Strategic Management Journal, 14*(S2), 95–112.

Lewis, M. W. (2000). Exploring paradox: Toward a more comprehensive guide. *Academy of Management Review, 25*(4), 760–776.

Limerick, D., & Cunnington, B. (1993). *Managing the new organisation*. Sydney: Business and Professional Publishing.

Lowendahl, B. R., & Revang, O. (2004). Achieving results in an after modern context: Thoughts on the role of strategizing and organizing. *European Management Review, 1*(1), 49–54.

Lubatkin, M. H., Simsek, Z., Ling, Y., & Veiga, J. F. (2006). Ambidexterity and performance in small-to medium-sized firms: The pivotal role of top management team behavioral integration. *Journal of Management, 32*(5), 646–672.

March, J. G. (1991). Exploration and exploitation in organizational learning. *Organization Science, 2*(1), 71–87.

Naisbitt, J., & Aburdene, P. (1990). *Megatrends 2000: Ten new directions for the 1990's*. New York: Avon Books.

Nosella, A., Cantarello, S., & Filippini, R. (2012). The intellectual structure of organizational ambidexterity: A bibliographic investigation into the state of the art. *Strategic Organization, 10*(4), 450–464.

O'Reilly, C. A., & Tushman, M. L. (2004). The ambidextrous organization. *Harvard Business Review, 82*(4), 74–81.

O'Reilly, C. A., & Tushman, M. L. (2008). Ambidexterity as a dynamic capability: Resolving the innovator's dilemma. *Research in Organizational Behavior, 28*(1), 185–206.

Palmer, I., & Dunford, R. (2002). Out with the old and in with the new? The relationship between traditional and new organizational practices. *International Journal of Organizational Analysis, 10*(3), 209–226.

Palmer, I., Benveniste, J., & Dunford, R. (2007). New organizational forms: Towards a generative dialogue. *Organization Studies, 28*(12), 1829–1847.

Pascale, R. (1990). *Managing on the edge: How successful companies use conflict to stay ahead. New York: Viking Penguin*.

Pettigrew, A. M., & Fenton, E. M. (2000). Complexities and dualities in innovative forms of organizing. In A. M. Pettigrew & E. M. Fenton (Eds.), *The innovative organization* (pp. 279–300). London, UK: Sage.

Pettigrew, A. M., Whittington, R. L., Melin, L., Sanchez-Runde, C., Van Den Bosch, F. A. J., Ruigrok, W., et al. (2003). *Innovative forms of organizing*. London, UK: Sage.

Quinn, R. E., & Cameron, K. S. (Eds.). (1988). *Paradox and transformation: Toward a theory of change in organization and management*. Cambridge, MA: Ballinger Publishing.

Raisch, S., Birkinshaw, J., Probst, G., & Tushman, M. L. (2009). Organizational ambidexterity: Balancing exploitation and exploration for sustained performance. *Organization Science, 20*(4), 685–695.

Raynor, M. E., & Bower, J. L. (2001). Lead from the center: How to manage divisions dynamically. *Harvard Business Review, 79*(5), 93–100.

Smith, W. K., & Tushman, M. L. (2005). Managing strategic contradictions: A top management model for managing innovation streams. *Organization Science, 16*(5), 522–536.

Turner, N., Swart, J., & Maylor, H. (2013). Mechanisms for managing ambidexterity: A review and research agenda. *International Journal of Management Reviews, 15*(3), 317–332.

Tushman, M. L., & O'Reilly, C. A. (1996). The ambidextrous organization: Managing evolutionary and revolutionary change. *California Management Review, 38*(4), 8–30.

Tushman, M., Smith, W. K., Chapman Wood, R., Westerman, G., & O'Reilly, C. A. (2010). Organizational designs and innovation streams. *Industrial and Corporate Change, 19*(5), 1331–1366.

Ulrich, D., & Wiersema, M. F. (1989). Gaining strategic and organizational capability in a turbulent business environment. *The Academy of Management Executive, 3*(2), 115–122.

Volberda, H. W. (1998). *Building the flexible firm*. Oxford: Oxford University Press.

Voss, G. B., & Voss, Z. G. (2013). Strategic ambidexterity in small and medium-sized enterprises: Implementation exploration and exploitation in produce and market domains. *Organization Science, 24*(5), 1459–1477.

Wang, C., & Ahmed, P. (2003). Structure and structural dimensions for knowledge-based organizations. *Measuring Business Excellence, 7*(1), 51–62.

Weick, K. E. (1979). *The social psychology of organization* (2nd ed.). Reading, MA: Addison-Wesley.

CHAPTER 3

Duality Theory

Abstract This chapter argues that duality theory offers the greatest scope as a conceptual framework for connecting ambidexterity capabilities and simultaneous explore—exploit outcomes. Duality theory provides direction by emphasizing characteristics such as dynamism, minimal thresholds, and improvisation. Collectively, these elements engender responsive, adaptive thinking across interconnected explorative and exploitative ventures. The chapter includes a comprehensive review of duality theory and its evolution. Drawing on this review and ensuing critique of duality characteristics, it maintains that ambidexterity capability underpinned by five duality characteristics reinforces the organizing tension that delivers both explore and exploit outcomes. The chapter concludes by proposing developmental measures for enhancing ambidexterity capabilities.

Keywords Duality theory · Improvisation · Enhancing ambidexterity

INTRODUCTION

As noted in Chap. 2, harnessing the explore—exploit tension to create ambidextrous forms of organizing leads to the successful management of continuity and change. Yet to date, no comprehensive theory exists to seamlessly guide organizations toward ambidexterity capabilities. Although we all seem to agree that innovation and control remain relevant, achieving both in high concentrations at the same time has so far relied as much on

© The Author(s) 2017
A.C.T. Smith et al., *Reinventing Innovation*,
DOI 10.1007/978-3-319-57213-0_3

trial and error as on sound theoretical precepts. We do, however, have some useful starting points to help.

To begin with, as complex, dynamic, and open systems, organizations need to learn to mobilize rather than suppress duality tensions and allow the complementarities to flourish (Hampden-Turner 1990a, b; Lewis 2000). Building ambidexterity capability equips organizations with comfort about tension, leading to a natural acceptance of explore and exploit initiatives that might conventionally be viewed as contradictory. In this chapter, we argue that duality theory offers the greatest scope as a conceptual framework for connecting ambidexterity capabilities and simultaneous explore—exploit outcomes. Duality theory provides direction by emphasizing characteristics such as dynamism, minimal thresholds, and improvisation. Collectively, these elements engender responsive, adaptive thinking and interconnectivity across explorative and exploitative ventures. Duality characteristics are explored further shortly.

Next, we begin with a comprehensive review of duality theory and its evolution. Drawing on this review and ensuing critique of duality characteristics, we argue that ambidexterity capability, underpinned by the five duality characteristics, reinforces an organizing tension that delivers both explore and exploit outcomes. Specifically, we propose duality theory as an explanatory framework and outline five indicative developmental measures for enhancing ambidexterity capabilities.

Duality Theory: An Explanatory Framework

Duality theory is different from, but related to, dualism. Dualism is a long-enduring approach to classification where an object of study is divided into paired and opposite elements, the most common examples being mind/body and theory/practice (Jackson 1999). Dualism represents an unambiguous, clearly demarcated contrast. Duality theory on the other hand—originally a by-product of Giddens' (1984) structuration theory—maintains the dualism belief that there are two essential components, but views them as interdependent (complementary *and* contradictory), rather than separate and opposed (Farjoun 2010). As with structuration theory, duality theory recognizes the complex interplay between agency and structure (Reed 1997) and the fact that 'agency derives from the simultaneously enabling and constraining nature of the structural principles by which people act' (Whittington 1994, p. 72). Duality theory thus offers a robust framework for viewing exploration and exploitation as distinctive, yet complementary components. Instead of being opposite sides of the same coin, exploration and exploitation are better understood as components in a mixed drink, or a recipe, where the absence of one component leads to diminution of the whole.

The problem arises when an organization chooses to focus on one of the explore—exploit duality poles at the expense of the other, thus creating 'a tension and difficulty to enact both ends of the continuum simultaneously.' Evans and Doz (1992, pp. 256–257) underlined the problem, arguing that 'almost all qualities of an organization have a complementary opposite quality, and excessive focus on one pole of a duality ultimately leads an organization into stagnation and decline (undue continuity), while the corrective swing to the opposite pole leads to disruptive and discontinuous crisis (excessive change).'

Dualities are not simply alternatives (Seo et al. 2004, p. 74). Firm growth through *exploration*—encouraging creativity and risk-taking, and tapping into new, untested markets and opportunities—can only occur from a stable *exploitation* foundation, concerned with accountability and control, and the capacity to increase efficiency and profitability. It is necessarily a dual-pronged approach, involving the exploration of new possibilities and the exploitation of old certainties (March 1991). While there may be trade-offs between the two, maintaining an appropriate balance is 'critical for firm survival and prosperity' (March 1991, p. 482). A more holistic way of thinking should highlight the complementarities rather than the contradictions. Explore—exploit dualities operate as interdependent, bimodal work practices (Fenton and Pettigrew 2000; Whittington and Pettigrew 2003). However, as we have argued, the ability to work with, and benefit from, the explore—exploit tension depends upon building ambidexterity capability.

Interest in the role dualities play in shaping structure and action in organizations gained urgency as leaders sought innovative, more responsive forms of organizing in the face of increasing environmental ambiguity. As Lewin et al. (1999, p. 541) observed about robust (dynamic) systems, a constructive tension emerges between order (the push of exploitation) and disorder (the pull of exploration). Similarly, Graetz and Smith (2008) highlighted the same tensile disjunction between the need for efficiency and stability through exploitation of existing resources, and opportunistic responsiveness through exploration and experimentation. Building on earlier studies of dualities and duality theory (Cameron 1986; Evans and Doz 1989, 1992; Quinn and Cameron 1988; Weick 1982), Graetz and

Table 3.1 Explanation of duality characteristics

Simultaneity	Represents the simultaneous presence of contradictory forces and starting point for understanding and managing organizational dualities. Captures the heterogeneous, qualitative nature of dualities
Relatedness	Represents the bidirectional, interdependent relationship between opposite poles. Highlights the advantages that come from managing contradictions as complementarities. Also illustrates the importance of both/and mindset facilitated through dynamic balance of minimal thresholds
Minimal thresholds	Maintains a dynamic balance between enabling and constraining forces to avoid strategic inertia and ensure organization remains poised on the competitive cusp between order and disorder
Dynamism	Dynamism (infused with simultaneity, relatedness, and minimal thresholds) ensures ongoing flexibility, creativity, and adaptability by encouraging dynamic interaction between duality poles
Improvisation	Represents a fusion of intended and emergent action, drawing on simultaneity, relatedness, minimal threshold, and dynamism to mediate between constantly contradictory goals. Works deliberately and extemporaneously to ensure that contradictions become complementarities

Smith (2008) developed five duality characteristics to assist leaders and managers work toward dualities. These five duality characteristics were based on key themes that emerged from a detailed synthesis of the literature on organizing forms. The five duality characteristics were: (1) simultaneity; (2) relational; (3) minimal thresholds; (4) dynamism; and (5) improvisation (summarized in Table 3.1). The following critique also considers the application of dualities in identifying critical explore—exploit tensions. In turn, it becomes more transparent how ambidexterity capabilities can be built.

DUALITY CHARACTERISTICS

Identifying duality characteristics enables us to consider the components of a productive organizing architecture. The right design accommodates contradictions by seeking to leverage, rather than reconcile or switch between, exploitative continuity activities and explorative change activities. Our approach rests on the premise that an ambidexterity capacity for adaptive, 'both/and' thinking that involves key people at multiple organizational levels is 'central to creating and sustaining competitive advantage' (Liedtka 1998, p. 31). Below we describe each duality characteristic based on the Graetz and Smith (2008) taxonomy.

Simultaneity. Graetz and Smith (2008) argued that simultaneity captures the heterogeneous, qualitative nature of dualities. It represents the foundational *starting* point for understanding and managing organizational dualities because it locates the presence of explorative and exploitative organizing elements within a single structural reference frame. This takes into account Evans and Doz' (1992) dualistic principle that all forms of social systems have *simultaneous* needs that appear contradictory but are in fact complementary. For example, top-down management processes must be combined with bottom-up processes. In addition, adaptation, learning, and innovation all demand loose, organic properties while efficiency and profitability demand tight, mechanistic qualities. Bimodal thinking relies on simultaneity, at the same time mitigating the risk of favoring either exploration or exploitation at the expense of the other.

Relatedness. The 'relational' characteristic relatedness points to the interactive, symbiotic attributes of the dualistic tensions that arise from the simultaneous pursuit of exploitation and exploration. According to Graetz and Smith (2008), the relational, interdependent nature of dualities suggests that organizational forms are not separate from the other organizational practices they house. The co-dependence of the explore—exploit meta-duality was also reinforced by Seo et al. (2004). They described dualistic relationships as bipolar, implying multiple interrelated tensions. For example, certainty-uncertainty can become entwined with expected-unexpected and routine-non-routine in a larger system of bipolar pairs. In addition, relational interdependence within these dual forms of organizing highlights the advantages that come from managing contradictions as complementarities (Whittington and Pettigrew 2003). Such interrelated, intersecting tensions are representative of the competing, yet complementary 'pairs' that comprise exploration and exploitation. Other examples include responsiveness/control, freedom/accountability, intuition/analysis, and ends/means.

Minimal threshold. We return to the importance of a 'both/and' rather than 'either/or' mindset. In order to yield a 'both/and' outcome, organizations need to reach a minimal level of explore and exploit activity. The minimal threshold characteristic thus serves to circumvent polarization where one organizing domain is favored at the expense of the other. Evans and Doz' (1992) third dualistic principle argued for maintaining minimal thresholds to avoid strategic inertia and ensure that an organization has the capacity for both change and continuity. Furthermore, they claimed, a focus on one polarity results in organizational instability and disintegration. As Hedberg et al. (1976) advocated, organizations should maintain a minimal threshold of desirable attributes; there should be minimal consensus to ensure the status quo is not implacable and unchallenged. They advocated minimalism in all forms of organizing to ensure a *dynamic* balance between creative responsiveness and disciplined efficiency. For example, on the exploitation side, there needs to be just enough, but not too much hierarchy, rationalization, specialization, centralization, and so on, thereby maximizing the push-pull tension between continuity and change. Organizations want unfettered innovation of course, but it also has to remain cost controlled and relevant.

Applying the minimal threshold 'test' to both exploration and exploitation activities help identify excessive 'swings' in one direction or the other. One common example comes in the dangerous form of 'bureaucracy-creep' for small, entrepreneurial start-up organizations as growth leads to swelling formalization and control, structural complexity, greater specialization, and the imposition of more and more rules and regulations. When the minimal threshold duality characteristic and its alliance with relatedness and simultaneity is understood and applied, it signals the need for a correction. An intervention can then proceed, in so doing obviating the adverse impact of pursuing exploitation capabilities at the expense of the organization's explorative, entrepreneurial activities that brought it success in the first place.

Dynamism. By highlighting the interdependent, complementary nature of dualities, the relatedness characteristic also reveals the dynamic and flexible quality of dualities, with constant interplay occurring between explore and exploit. Dynamism presents an important consideration in the context of environmental turbulence as it presupposes that survival depends on both exploration and exploitation. By encouraging a reciprocal tension between exploration and exploitation, the characteristic 'dynamism' keeps the characteristic 'minimal thresholds' in tune to ensure ongoing flexibility, creativity, and adaptability. Ongoing dialogues between explore and exploit thinking alerts organizations to environmental shifts and pressures, in turn highlighting the characteristic 'improvisation' in managing the push-pull tension.

Improvisation. The fourth characteristic 'dynamism' underlines not only the vital flow between the dual explore—exploit organizing poles, but also the evolving, active nature of organi*zing* as opposed to organi*zation*.

Coping with complexity and ambiguity, and being able to adapt and respond proactively to environmental shifts, is particularly important as organizations establish external partnerships and an international focus. The 'improvisation' characteristic can play an important role here as the medium for interaction across the other four characteristics. Weick (1998, p. 552) described improvisation as a fusion of intended and emergent action that manifests as 'a mixture of the pre-composed and the spontaneous.' He maintained that by recognizing the overarching, mediating role of improvisation, organizations were more likely to appreciate the importance of both intended and emergent actions, and avoid the temptation of favoring one over the other.

Improvisation arbitrates between organizational dualities through 'adjustments.' In practice, this means tinkering with organizing elements rather than by trying to resolve what seem to be contradictions (Graetz and Smith 2006). Improvisation serves as the linchpin between the two poles of 'plans' and 'action.' Plans get amended through improvisation as the result of changing circumstances, and then enacted (Clegg et al. 2002, p. 492). As a consequence, improvisation delivers the pivotal medium through which dynamism, simultaneity, minimalism, and relatedness are created, revised, and discovered (Weick 1998).

DUALITY CHARACTERISTICS AND AMBIDEXTERITY CAPABILITY

We propose that ambidexterity capability, underpinned by proficiency with the five duality characteristics, helps build connections between competing explore—exploit tensions such as individual/organizational, static/dynamic, internal/external, and differentiation/integration. For example, activating internal/external and differentiation/integration tensions through the duality characteristics dynamism, simultaneity, and relatedness enables an organization to sense and seize (Teece 2007) opportunities. The result can stimulate a reconfiguration of business systems and structures. When high-tension connections are transformed into accepted organizing routines through ambidexterity capability, organization members become accustomed to dealing with novelty and ambiguity. Thinking laterally with duality characteristics can result in a superior diffusion of knowledge and information across explorative and exploitative processes. We now outline five potential measures that may provide the means for nurturing and encouraging ambidexterity capabilities leading to explore and exploit success.

FIVE POTENTIAL MEASURES FOR BUILDING AMBIDEXTERITY CAPABILITY

Figure 3.1 outlines the potential intervention measures and contextualizes these against the interdependent explore—exploit goals, and the corresponding duality characteristics that underscore their realization.

Measure 1. An Effective and Committed Leadership Team with Dualities Awareness

A first primary condition to building ambidexterity capability is a committed leadership team with the imperative to act as key connectors and boundary spanners, facilitating the flow of information between different groups and enabling 'communication and understanding to take place across ... knowledge domains' (Taylor and Helfat 2009, p. 721). Brooks and Saltzman (2016, p. 64), for instance, referred to the central role of boundary spanners who enable organization members to 'cross-pollinate, share resources, and stay calibrated with the most important stakeholders.' Case after case on organizational change reveals that collaborative and synergistic activities drive success. At the core remain skilled leaders who can think and act 'in broad and integrative ways' (Chen and Miller 2010, p. 9; Doz and Thanheiser 1993; Kotter 1995; Nadler et al. 1995; Paine 2010). For example, Turner et al. (2016, p. 214) proposed that integrative actions represent the 'spine' of project-based ambidexterity. As captured in the interrelated features of the exploit—explore goals, individual/organizational, static/dynamic, and leadership/empowerment, it becomes clear that leaders must direct as well as empower—to be decisive and exert authority alongside encouraging teamwork and individual initiative.

Early studies on ambidextrous forms of organizing focused on structural ambidexterity, arguing the case for a separation of explorative and exploitative business units (He and Wong 2004; O'Reilly and Tushman 2004; Tushman and O'Reilly 1996). In contrast, Gibson and Birkinshaw (2004) investigated the potential of contextual ambidexterity to manage the explore—exploit organizing tension. The researchers determined that contextual ambidexterity involves the human, behavioral side of the organization and the choices people make between alignment-focused and adaptability-focused activities (Turner et al. 2016). Similarly, in their study of the role of top management teams and leadership behavior in achieving ambidexterity, Jansen et al. (2009) emphasized the importance of

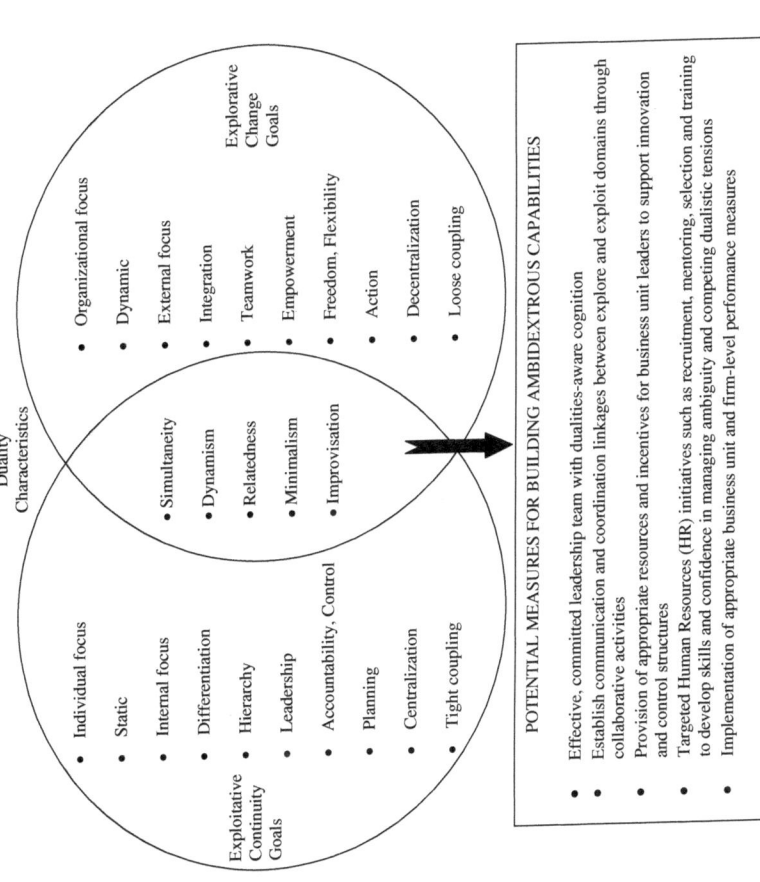

Fig. 3.1 Conditions for exploitative continuity and explorative change: building ambidexterity capability

behavioral-oriented integration mechanisms, such as social integration among senior team members and cross-functional interfaces for knowledge and skills sharing across explore and exploit domains. Turner et al. (2016, p. 215) further substantiated the importance of contextual factors and identified five managerial actions that contributed to the achievement of what they considered ambidextrous balance. These included: 'role expansion' (taking ownership of problems); 'tone setting' (framing and modeling the explore—exploit mindset); 'buffering' (providing managerial 'protection' from outside intervention or distraction); 'gap filling' (performing administrative tasks often overlooked); and 'integration' (actively bringing together diverse skills and knowledge).

The previously outlined findings (Gibson and Birkinshaw 2004; Jansen et al. 2009; Turner et al. 2016) support our contention that in order to build ambidexterity capability, a leadership team must first accept that they are working with dual organizing architectures and that both are integral to performance and growth. Moreover, leaders must be unified in articulating their mission and the values that underpin ambidexterity capability. Leadership must be able to engage, involve, and link key stakeholders across both explore and exploit domains. Such ambidexterity capability sponsors bring together organization members with divergent skills, knowledge, and expertise. By building trust and respect for diversity, key stakeholders expand the depth and breadth of ambidexterity capability.

In considering the characteristics of exploitation, we appreciate that hierarchy, in alliance with tight coupling, accountability, and control, provides structure and guidance, demarcating roles, responsibilities, and reporting relationships. But we have also learned that the dual organizing tension demands an optimization of efficiency and flexibility. Hierarchy, therefore, should not arrive at the expense of delayering principles that mitigate against overly tight structures that lead to rigidity, excessive reporting, and heavy-handed, inertial bureaucracy. However, at the other extreme, abandoning hierarchy in favor of loose, flat structures risks moving from loose coupling to decoupling, and creating a rudderless organization lacking direction or control. Supporting this view, Paine's (2010) findings illustrated the value of recognizing the dualistic tensions that stimulate a healthy, interactive empowerment-leadership dynamic. According to Paine, the most effective leaders were: strategic *and* hands-on; disciplined *and* entrepreneurial; concerned with process as well as people; combined firm, authoritative leadership with a more facilitative approach; and were willing to take calculated risks while being equally

cautious. Keller and Weibler (2015) also highlighted the importance of a balanced behavior repertoire, seen as essential for ambidextrous managers who must 'match' their engagement in exploration and exploitation. Developing an effective and committed leadership team illustrates the interplay of duality characteristics such as dynamism and relatedness, and the central role managers must play in creating linkages. The complexities of leadership action and behavior also underline the importance of maintaining a *sufficient* (minimal) constructive tension between each end of the continuum that serves to stimulate rather than constrain change and action. In addition, establishing connections through collaborative, cross-functional activities helps build comfort with diversity and ambiguity, reframing team identity and teamwork (Janssens and Steyaert 1999; Limerick and Cunnington 1993; Morgan 1997). Connections are closely allied with the following measure to establish effective communication and coordination linkages.

Measure 2. Communication and Coordination Linkages Between Explore and Exploit Domains

Allied to an effective and committed leadership, communication from senior management comprises a powerful lever in gaining acceptance and building consensus toward new ways of working (Jackson 2000; Kotter 1995; Kouzes and Posner 1995; O'Reilly and Tushman 2004). The placement of skilled, committed leaders and change advocates also bolsters effective communication and coordination linkages, effectively fostering the dialogue between explore and exploit domains (Taylor and Helfat 2009). As a number of studies have found, connections build social capital and support revised patterns of action and behavior (Gibson and Birkinshaw 2004; Lubatkin et al. 2006; Turner and Rindova 2012). For example, Gibson and Birkinshaw's (2004) behavior-framing attributes of support and trust, reflecting members' willingness to share information and expertise, and mutual confidence in each other's ability, proved critical to ensuring an open and supportive networking environment.

Coordination and communication linkages rely on leaders and managers to encourage relatedness, dynamism, and simultaneity. In addition, minimal thresholds ensure fluidity across boundaries to facilitate collaboration and information sharing, while improvisation fosters adaptive, proactive action in response to the unexpected.

Measure 3. Resources and Incentives for Business Unit Leaders to Support Innovation and Control Structures

If exploration and exploitation are to operate in active tension, business unit leaders accustomed to working in traditional exploit-based business units need to develop confidence in working with the ambiguity inherent in dual architectures. They must be equipped with the knowledge and communicative ability to explain new initiatives to business unit members, not least how they will be measured and rewarded. They also bear responsibility for developing business unit confidence, trust, and the ability to work in cross-functional, collaborative projects both internal and external to the organization. As Jansen et al. (2008, p. 999) found, achieving organizational ambidexterity demanded not only 'a strong and compelling shared vision,' but also 'shared fate' contingency rewards to ensure a combined explore—exploit approach.

Leaders need to illustrate how the innovations and adaptations arising from these shared initiatives will not only help the business unit grow its portfolio, but also enrich unit member knowledge and expertise. These are complex requirements that demand significant investment over the long term. The complexity and interwoven nature of these requirements signal that relatedness, dynamism, and improvisation act as key mediating levers.

Measure 4. Targeted Human Resources (HR) Initiatives (such as Recruitment, Mentoring, Selection, and Training) to Develop Skills and Confidence in Managing Ambiguity and Competing Dualistic Tensions

Another high-potential focal point concerns whether targeted human resources (HR) initiatives can play a role in developing the skills to manage competing dualistic tensions. Swart et al. (2016), for example, concluded that human resource management practices supporting integration played a key role in enabling senior managers to embrace both exploration and exploitation. Underlining the explorative and exploitative characteristics represented in Fig. 3.1, Raisch et al. (2009) observed that the key requirements for dealing with complex, uncertain environments include access to effective top-down/bottom-up and lateral knowledge flows, a capacity to analyze and interpret both short- and long-term issues, and a robust prior-related knowledge and experience base that fuels curiosity and facilitates new knowledge. We suggest that dualistic attributes need to be factored into mentoring and leadership training programs that work outward from the exploitative continuity and explorative change activities. The key lies in developing duality enabling work practices that underpin ambidexterity capability.

Some examples are instructive. Consider the traditional operations of an established business unit (such as Accounting and Taxation), which are founded in traditional, exploitative continuity processes. The 'individual' exploitative characteristic is reflected in billable hours, fee for service clients, and traditional key performance measures. Standardization and mandated control processes to ensure consistency of service delivery and legal compliance highlight the essential conservatism of the exploitative 'static' characteristic. Similarly, compulsory professional accreditation and specialist knowledge invoke 'differentiation.' The core aspects of these exploitative processes, reflected in individualism, standardization and conservatism, and high specialization, thus conspire to invoke an 'internal,' closed shop, focus. While these may be necessary conditions for effective accounting and taxation service delivery, a duality framework provides the inspiration to expand the scope and substance of the offerings. In establishing ambidexterity capability, the role of explorative characteristics, such as 'organizational,' 'dynamic,' 'integration,' and 'external,' can be captured alongside the exploitative requirements of the Accounting and Taxation unit. For example, by working with IT and marketing experts, the accounting and taxation specialists can explore more innovative client service offerings, such as adding a digital component to an existing service in order to improve flexibility and convenience without compromising standards. In this way, 'organizational,' through the duality characteristics simultaneity, dynamism and improvisation, opens up 'individual' to teamwork and information sharing. The explorative characteristic, 'dynamic,' loosens the reins of 'static,' easing but not denying the need for conservatism, initiating more flexible, adaptive forms of organizing. 'Integration,' 'teamwork,' 'decentralization,' and 'flexibility,' mediated by the duality characteristics, inform 'differentiation,' 'hierarchy,' and 'control,' encouraging collaboration through the development of a shared services model to offer a distinctive, valued added client experience. The explorative characteristic 'external' broadens the 'internal' horizons, highlighting the importance of engaging with clients and partners, working not only to meet, but also exceed client needs.

Measure 5. Implementation of Appropriate Business Unit and Firm-Level Performance Measures

Operating across exploit—explore domains means that the traditional, singular concern for structure must evolve into a dual concern for structure and process. That means dual benchmarks and measures must exist side by side in order to manage the relationships between the two essential domains (Pettigrew and Fenton 2000). An important feature of the static-dynamic tension, relating to key performance measures, also serves to fix the spotlight on how the explore and exploit activities of a firm require a different set of management skills and focus.

On the exploit side of an organization, traditional business units are generally evaluated against standard key performance indicators (KPIs) and balance sheet criteria, including gross revenues, margins, profit and loss, and other control and efficiency-related metrics. The central focus of an explore unit, on the other hand, is on nurturing explorative ventures in their infancy with little current yield, but with potential for future growth and profit. However, explore units must also meet key performance indicators. They face significant pressure to demonstrate value, an imperative exacerbated by the absence of tangible metrics or balance sheet credits around exploration that other business unit leaders recognize and use to measure performance in their exploit-based business units. Developing and building appropriate, dual performance measurement mechanisms is an important intervention measure that must be tackled early to ensure organization-wide acceptance and commitment to both exploration and exploitation. Transparency in consultation and negotiation is vital to ensure appropriate reward and performance measures are instituted, and invoking relatedness, dynamism and improvisation could be helpful in gaining acceptance and commitment to these measures.

CONCLUSION

Since the 1960s, scholars have laid down the gauntlet to received management wisdom by unpacking the concept of organizational dualities and their role in managing the interplay between continuity and change. Today, the insight is revealed in the serious challenge organizations face in meeting the coexisting imperatives of exploiting current products and services while exploring potential products and service experiences. Rather than being displaced, the demand to meet multiple but often inconsistent contextual demands has become more forceful with time (Benner and Tushman 2003; Tushman and O'Reilly 2008).

The plea to organizations has been to put in place systems that can cope with ambiguity, ambivalence, and contradiction. Organizations must learn to manage the dualistic tensions that underpin exploration and exploitation. Examples include: nurturing innovation alongside rigorous financial and operational systems; fostering empowerment through strong and supportive leadership; considering the impact of economic realities on social goals; and balancing formalized, central controls and policies with decentralized decision making that would support more flexible forms of organizing. Achieving such aims demands that leaders and managers wield the duality characteristics of simultaneity, relatedness, dynamism, minimal thresholds, and improvisation. Duality theory offers a potential explanatory framework for understanding and managing the tensions that arise in building a capacity for simultaneous, dynamic, related exploration and exploitation through the medium of organizational ambidexterity. A duality theory framework provides a conceptualization of change incorporating complexity and contradiction, without the implicit emphasis on removing, micromanaging, ignoring, or denying tension or contradictions.

As argued in this chapter and the previous chapter, the key to pursuing exploration and exploitation with equal skill and fervor lies in addressing apparently competing yet ultimately complementary contradictions as natural, irreducible elements of organizational ambidexterity capability. Ambidexterity capabilities operate across explore and exploit domains and broker linkages to share, absorb, and translate into action the accumulated knowledge, skills, expertise, and experience residing in both. As Tushman et al. (2010, p. 1336) claimed, positive performance outcomes depend upon ambidextrous designs that encapsulate *heterogeneity* through multiple integrated architectures that are 'highly differentiated and inconsistent.'

We proposed five potential measures for building ambidexterity capability. These measures are designed to capture and leverage the dual exploit —explore tensions through: critical integrative activities such as appointing skilled business unit leaders; establishing critical communication and coordination linkages; establishing appropriate resources and incentives for business unit leaders to support innovation and control systems and structures; putting in place targeted human resources initiatives; and introducing appropriate performance measures that recognize and reward both explore and exploit initiatives. In the following chapter, we take duality theory into the practical domain and show how it can be enabled in an organization.

REFERENCES

Benner, M. J., & Tushman, M. L. (2003). Exploitation, exploration, and process management: The productivity dilemma revisited. *Academy of Management Review, 28*(2), 238–256.

Brooks, S. M., & Saltzman, J. M. (2016). *Creating the vital organization.* New York: Palgrave Macmillan.

Cameron, K. S. (1986). Effectiveness as paradox: Consensus and conflict in conceptions of organizational effectiveness. *Management Science, 32*(5), 539–553.

Chen, M. J., & Miller, D. (2010). West meets east: Toward an ambicultural approach to management. *Academy of Management Perspectives, 24*(1), 17–22.

Clegg, S. R., da Cunha, J. V., & e Cunha, M. P. (2002). Management paradoxes: A relational view. *Human Relations, 55*(5), 483–503.

Doz, Y., & Thanheiser, H. (1993). Regaining competitiveness: A process of organizational renewal. In J. Hendry, G. Johnson, & J. Newton (Eds.), *Strategic thinking: Leadership and the management of change* (pp. 293–310). Chichester, UK: Wiley.

Evans, P., & Doz, Y. (1989). The dualistic organization. In P. Evans, Y. Doz, & A. Laurent (Eds.), *Human resource management in international firms: Change, globalization, innovation* (pp. 219–242). London, UK: Macmillan.

Evans, P., & Doz, Y. (1992). Dualities: A paradigm for human resource and organizational development in complex multinationals. In V. Pucik, N. Tichy, & C. Barnett (Eds.), *Globalizing management: Creating and leading the competitive organization* (pp. 85–106). New York: Wiley.

Fenton, E., & Pettigrew, A. (2000). *Theoretical perspectives on innovative forms of organizing.* California: Sage.

Gibson, C. B., & Birkinshaw, J. (2004). The antecedents, consequences, and mediating role of organizational ambidexterity. *Academy of Management Journal, 47*(2), 209–226.

Giddens, A. (1984). *The constitution of society: Outline of the theory of structuration.* California: University of California Press.

Graetz, F., & Smith, A. (2006). Critical perspectives on the evolution of new forms of organising. *International Journal of Strategic Change Management, 1*(1–2), 127–142.

Graetz, F., & Smith, A. (2008). The role of dualities in arbitrating continuity and change in forms of organizing. *International Journal of Management Reviews, 10*(3), 265–280.

Hampden-Turner, C. M. (1990a). *Charting the corporate mind: From dilemma to strategy.* Oxford: Blackwell.

Hampden-Turner, C. M. (1990b). *Corporate culture: From vicious circles to virtuous circles.* London: Hutchinson/Economist Books.

He, Z. L., & Wong, P. K. (2004). Exploration vs. exploitation: An empirical test of the ambidexterity hypothesis. *Organization Science, 15*(4), 481–494.

Hedberg, B., Nystrom, P., & Starbuck, W. H. (1976). Camping on seesaws: Prescriptions for a self designing organization. *Administrative Science Quarterly, 21*(1), 41–65.

Jackson, D. (2000). *Becoming dynamic: Creating and sustaining the dynamic organization.* London, UK: Macmillan Business.

Jackson, W. A. (1999). Dualism, duality and the complexity of economic institutions. *International Journal of Social Economics, 26*(4), 545–558.

Jansen, J. J. P., George, G., Van den Bosch, F. A. J., & Volberda, H. W. (2008). Senior team attributes and organizational ambidexterity: The moderating role of transformational leadership. *Journal of Management Studies, 45*(5), 982–1007.

Jansen, J. J. P., Tempelaar, M. P., Van den Bosch, F. A. J., & Volberda, H. W. (2009). Structural differentiation and ambidexterity: The mediating role of integration mechanisms. *Organization Science, 20*(4), 797–811.

Janssens, M., & Steyaert, C. (1999). The world in two and a third way out? The concept of duality in organization theory and practice. *Scandinavian Journal of Management, 15*(2), 121–139.

Keller, T., & Weibler, J. (2015). What it takes and costs to be an ambidextrous manager: Linking leadership and cognitive strain to balancing exploration and exploitation. *Journal of Leadership and Organizational Studies, 22*(1), 54–71.

Kotter, J. P. (1995). Leading change: Why transformation efforts fail. *Harvard Business Review, 73*(2), 59–67.

Kouzes, J. M., & Posner, B. Z. (1995). *The leadership challenge.* San Francisco: Jossey-Bass.

Lewis, M. W. (2000). Exploring paradox: Toward a more comprehensive guide. *Academy of Management Review, 25*(4), 760–776.

Liedtka, J. M. (1998). Linking strategic thinking with strategic planning. *Strategy and Leadership, 26*(4), 30–35.

Limerick, D., & Cunnington, B. (1993). *Managing the new organisation.* Sydney: Business and Professional Publishing.

Lewin, A.Y., Long, C., & Carroll, T. (1999). The coevolution of new organizational forms. *Organization Science, 10*(5), 535–50.

Lubatkin, M. H., Simsek, Z., Ling, Y., & Veiga, J. F. (2006). Ambidexterity and performance in small-to medium-sized firms: The pivotal role of top management team behavioral integration. *Journal of Management, 32*(5), 646–672.

March, J. G. (1991). Exploration and exploitation in organizational learning. *Organization Science, 2*(1), 71–87.

Morgan, G. (1997). *Images of organization*. Thousand Oaks, CA: Sage.

Nadler, D. A., Shaw, R. B., & Walton, A. E. (1995). *Discontinuous change: Leading organizational transformation*. San Francisco: Jossey Bass.

O'Reilly, C. A., & Tushman, M. L. (2004). The ambidextrous organization. *Harvard Business Review, 82*(4), 74–81.

Paine, L. (2010). The China rules. *Harvard Business Review, 88*(6), 103–108.

Pettigrew, A. M., & Fenton, E. M. (2000). Complexities and dualities in innovative forms of organizing. In A. M. Pettigrew & E. M. Fenton (Eds.), *The innovative organization* (pp. 279–300). London, UK: Sage.

Quinn, R. E., & Cameron, K. S. (Eds.). (1988). *Paradox and transformation: Toward a theory of change in organization and management*. Cambridge, MA: Ballinger Publishing.

Raisch, S., Birkinshaw, J., Probst, G., & Tushman, M. L. (2009). Organizational ambidexterity: Balancing exploitation and exploration for sustained performance. *Organization Science, 20*(4), 685–695.

Reed, M. I. (1997). In praise of duality and dualism: Rethinking agency and structure in organizational analysis. *Organization Studies, 18*(1), 21–42.

Seo, M. G., Putnam, L. L., & Bartunek, J. M. (2004). Dualities and tensions of planned organizational change. In M. S. Poole & A. H. Van de Ven (Eds.), *Handbook of organizational change and innovation* (pp. 73–107). Oxford: Oxford University Press.

Swart, J., Turner, N., van Rossenberg, Y., & Kinnie, N. (2016). Who does what in enabling ambidexterity? Individual actions and HRM practices. *The International Journal of Human Resource Management*, 1–28. doi:10.1080/09585192.2016.1254106.

Taylor, A., & Helfat, C. E. (2009). Organizational linkages for surviving technical change: Complementary assets, middle management, and ambidexterity. *Organization Science, 20*(4), 718–739.

Teece, D. J. (2007). Explicating dynamic capabilities: The nature and microfoundations of (sustainable) enterprise performance. *Strategic Management Journal, 28*(13), 1319–1350.

Turner, N., Swart, J., Maylor, H., & Antonacopoulou, E. (2016). Making it happen: How managerial actions enable project-based ambidexterity. *Management Learning, 47*(2), 199–222.

Turner, S. F., & Rindova, V. (2012). A balancing act: How organizations pursue consistency in routine functioning in the face of ongoing change. *Organization Science, 23*(1), 24–46.

Tushman, M. L., & O'Reilly, C. A. (1996). The ambidextrous organization: Managing evolutionary and revolutionary change. *California Management Review, 38*(4), 8–30.

Tushman, M., Smith, W. K., Chapman Wood, R., Westerman, G., & O'Reilly, C. A. (2010). Organizational designs and innovation streams. *Industrial and Corporate Change, 19*(5), 1331–1366.

Weick, K. E. (1982). Management of organizational change among loosely coupled elements. In P. S. Goodman & Associates (Eds.), *Change in organizations: New perspectives in theory, research and practice* (pp. 375–348). San Francisco: Jossey-Bass.

Weick, K. E. (1998). Improvisation as a mindset for organizational analysis. *Organization Science, 9*(5), 543–555.

Whittington, R. (1994). Sociological pluralism, institutions and managerial agency. In J. Hassard & M. Parker (Eds.), *Towards a new theory of organizations* (pp. 53–74). London: Routledge.

Whittington, R., & Pettigrew, A. M. (2003). Complementarities thinking. In A. M. Pettigrew, R. L. Whittington, L. Melin, C. Sanchez-Runde, F. A. J. Van Den Bosch, W. Ruigrok, & T. Numagami (Eds.), *Innovative forms of organizing* (pp. 125–132). London, UK: Sage.

CHAPTER 4

Embracing the Tension

Abstract This chapter furthers the book's proposition that at the heart of duality theory resides the explore—exploit problem, which is concerned with how firms can stimulate innovation for the future while maintaining a high return upon existing opportunities. It also addresses how to build ambidexterity capability. Based on longitudinal case study data, the chapter suggests that one productive method for developing ambidexterity capability involves pursuing a dual organizational identity embracing innovation and efficiency as mutually inclusive pursuits. The chapter draws on the case evidence to show that a leader sensegiving/sensemaking communications strategy needs to shift from one formulated around constancy, efficiency, and control to one imbued with duality thinking. The chapter also connects the case evidence with theory.

Keywords Case study · Sensegiving · Sensemaking · Identity

INTRODUCTION

This chapter furthers the practical enactment of our proposition that at the heart of duality theory resides the explore—exploit problem, which is concerned with how firms can stimulate innovation for the future while maintaining a high return upon existing opportunities. It also addresses the practical question of how to build the kind of ambidexterity capability upon which an innovation foundation can be grounded, while maintaining efficient, high-yield core business operations. Based on the longitudinal

© The Author(s) 2017   55
A.C.T. Smith et al., *Reinventing Innovation*,
DOI 10.1007/978-3-319-57213-0_4

case study data examined here, we suggest that one productive method for developing ambidexterity capability involves pursuing a dual organizational identity embracing innovation and efficiency as mutually inclusive pursuits. Of course, this sounds easy in theory, but the reality presents a far more messy and complex task. As a result, we take a deep dive into a genuine case where a company actually made it work. In the process of examining the case, we look carefully at all the obstacles and opportunities that arose.

In order to establish what appear to be contradictory goals, the case firm's leadership successfully implemented an identity transformation encouraging its constituents toward a dual explore—exploit organizing form. The chapter's case maps a change process dedicated to successfully marrying the predictable commercial returns of an existing but declining business with the uncertain pursuit of nascent but high-yield new business.

What makes this case unique were the tactics employed by executive leaders as they sought to introduce a more entrepreneurial, 'innovation' mindset alongside the traditional business model. The chapter exposes these ambidexterity building tactics by applying a sensegiving and sensemaking analysis—the shared beliefs and interpretations about organizational life—to the communication between the firm's executive leadership and its partners, directors, and senior managers. In so doing, our analysis shows how they successfully managed an identity transition. Such data represent a key step in extending duality theory from concept into potential application and testable hypotheses. It also sheds some light on how a firm can cultivate ambidextrous capability, the forerunner to the 'dual organization.'

The case firm has been in business in some form since the later part of the nineteenth century. It grew and prospered until the late 1980s when it became the victim of its own success, having become large but inert, lacking the drive, willingness, and flexibility to adapt and respond to a changing environment. At this time, the focus was on ensuring efficiency and stability, driven by conservative organizing principles around regulatory requirements and accountabilities. By the 1990s, the company had lost focus and direction, fueled by the arrival and departure of 10 CEOs in the space of just eight years. By 2003, as the newly appointed CEO discovered, the company was hemorrhaging clients, staff, and millions of dollars of revenue. As he described it, the place 'was just a mess, it was not a happy place, and it was not a happy commercial place,' lacking in motivation and low in morale and collective esteem. However, the incoming CEO and his new deputy, the firm's Chief Strategy Officer (CSO), saw the widespread turmoil and discontent as an opportunity to construct a new vision and direction. Their ambition was to encourage a more innovative, entrepreneurial business model to enhance the traditional side of their services that relied upon defending existing routines for business success. The new leadership sought to encourage opportunistic responsiveness through *exploration* and experimentation while simultaneously maintaining efficiency and stability through the *exploitation* of existing resources. Initially, however, the senior leadership team struggled to give their message a clear voice. As Sonenshein (2010, p. 504) noted, communicating 'interwoven progressive and stability' messages risks causing ambiguity and concern among organizational members, leading to heightened resistance while undermining the change process. It was not until the firm's leaders explicitly transformed their message and *consistently* sought to emphasize both explore and exploit endeavors as integral to the organization's continued success, that change arrived. Slowly, those middle managers and partners responsible for implementing the new vision began to understand what new behaviors and practices were required to build such ambidexterity capabilities.

To foreshadow the case findings, we identified four characteristic themes emerging from the firm's experiences: (1) shaping a dual (explore—exploit) organizational identity to embrace ambiguity and uncertainty; (2) enabling explore—exploit thinking, which helped to release and open new ways of looking at the traditional change problem; (3) the intentional aim to shape meaning through sensegiving and sensemaking, which strategically demonstrated how explore—exploit could be interpreted and understood through personal and practical exemplars; and (4) enabling dual explore—exploit endeavors by building ambidexterity capability, which provided tangible systems and structures to underpin the new way of working. The remainder of the chapter presents the background, research approach, data, results, and implications of the case. Where pertinent, we connect the practice with theory.

SENSEMAKING AND SENSEGIVING: MAKING MEANING IN SHAPING DUAL AMBIDEXTERITY CAPABILITIES

Every organization holds shared beliefs reflected in forms of communication that increase social cohesion and solidarity. In fact, organizational members share common beliefs and interpretations of their own experiences as well as of external events. Berger and Luckmann (1967) proposed that behavior in institutions comes about because certain actions are encouraged, reinforced, and repeated until patterned and habitualized. These behaviors become ingrained in both the performers' constructions of the institution's identity and the institution's collective activities through a multiplicity of individual actions. As a consequence, reality is legitimized through individuals' role-playing, leading to a shared, plausible, and meaningful social reality.

Of particular salience to this case is Berger and Luckmann's (1967) suggestion that 'legitimization' helps transmit already-established institutional meaning via systems of symbolic reference: 'Language constructs immense edifices of symbolic representation that appear to tower over the reality of everyday life like gigantic presences from another world' (Berger and Luckmann 1967, p. 55). Within this schema, sensemaking and sensegiving may be considered central to the 'social distribution of knowledge' (Berger and Luckmann 1967, p. 146).

Through their 'sensemaking' and 'sensegiving,' leaders are able to 'use the raw materials of narrative to construct new organizational sense' (Fleming 2001, p. 34). Gioia and Chittipeddi (1991) conceptualized the sensemaking/sensegiving process as pivotal to leadership communications in articulating a new organizational identity. Leader sensegiving represents a critically important activity in shaping 'the processes and outcomes of organizational sensemaking' (Maitlis and Lawrence 2007, p. 57), because it serves to guide organizational actors' interpretations and meaning making towards a new reality. Other studies have similarly commented on the importance of leader sensegiving in bringing about organizational change (Dunford and Jones 2000; Fiss and Zajac 2006; Maitlis and Lawrence 2007; Weick 1995). Through language, talk, and communication, 'meanings' materialize from sensegiving and sensemaking that 'inform and constrain identity and action' (Weick et al. 2005, p. 409). From a leadership perspective, the sensemaking/sensegiving dynamic means replacing old ways of looking at things with new ways (Gioia and Chittipeddi 1991).

Organizing forms literature suggests that top management plays a pivotal role in providing 'connective tissue' and in engendering social integration among senior team members and cross-functional interfaces for knowledge and skills sharing (Birkinshaw and Gibson 2004; Simsek et al. 2005; Turner et al. 2012). Leader sensegiving and sensemaking 'craft' can therefore play a significant part in shaping understandings about the organizational change process through the words, ideas, and images leaders employ (Peirano-Vejo and Stablein 2009). Accordingly, it is up to leaders to use persuasion to craft a new reality.

Acting as 'sensegiver' (Dunford and Jones 2000), a leader's subtle and strategic interpretation and contextualization of events, conveyed through the rhetoric of metaphor, stories, and repetition, serve to construct a compelling change message. In this respect, a leader's language can reshape organizational reality by delineating acceptable modes of social engagement. In times of uncertainty and change it remains particularly important for a leader to 'develop a vision or mental model of how the environment works (sensemaking) and then be able to communicate to others and gain their support (sensegiving)' (Hill and Levenhagen 1995, p. 1057).

CASE AND CONTEXT

Our initial research question emerged from the explore—exploit problem and the absence of case data revealing how it has been addressed in practice. The dualities concept offers a theoretical platform for conceptualizing the tensions that arise when firms seek efficiency and innovation simultaneously by responding to both at the same time (Smith and Lewis 2011). However, little empirical work—especially from cases—exposes how leaders can create an organizational identity where explore and exploit ambitions are simultaneously fostered. Our research aimed to bolster duality theory by providing some avenues for developing ambidexterity capabilities in practice.

We examined sensegiving and sensemaking communications through the lens of duality theory, the latter used to guide interpretation of the themes emanating from communications documented over a three-year process of data collection, including interviews and ethnography. A duality lens encourages a 'both/and' approach by exploring the links and synergies between the apparently opposing dimensions of exploration and exploitation. Consequently, examining the nature of sensegiving/sensemaking through a dualities lens provides insights into the ways in which the simultaneous pursuit of exploration and exploitation can be shaped and 'talked' into reality by senior leaders.

Our approach recognized the prior knowledge that duality theory offers about the explore—exploit research question, but did not employ predetermined hypotheses for testing against data. In considering grounded techniques of analysis, Strauss and Corbin (1990, p. 56) observed that, '... all kinds of literature can be used before a research study is begun...' We therefore positioned our analytical approach as interpretive, informed by a theoretical platform framing the research problem, but inductive in that a naturalistic inquiry method was used consistent with the method advocated by Lincoln and Guba (1985) in order to allow for an emergent presentation of the primary data. Accordingly, a naturalistic inquiry should outline its theoretical perspective. As a result, the naturalistic inquiry approach utilized in this research employed a conceptual coding method (Strauss and Corbin 1990), which was guided by the approach suggested by Denis et al. (2001). They outlined how nascent theory can be explored and developed by starting with a theory-inspired problem and pursuing it with a detailed inductive procedure. We began with the explore—exploit problem inspired by duality theory, and used conceptual coding to compile the data into themes relevant to the case firm's response to the problem. In practice, these themes represent the architecture around which the firm constructed its ambidexterity capability.

Case studies allow researchers to understand events and issues from the perspective of different organizational stakeholders (Gillham 2000), especially where people and structures intersect in complex ways. While case study methodology is sometimes criticized for its lack of generalizability, relevant to our approach was Yin's (2009, p. 15) argument that case study findings are generalizable to 'theoretical propositions,' not simply organizations or populations. This is supported by Farquhar's (2012) view that the primary intent of the case study is to conduct an in-depth study of a *single* case or multiple cases, rather than generalize findings. In addition, the evidence collected in cases emerges from the social setting being studied, strengthening its validity (Jennings 2001).

Data were collected through a series of detailed individual histories using in-depth interviews seeking multilayered responses. The interviews were conducted as conversations (Riessman 2008) where the participants were encouraged to provide an account of their sensemaking histories relevant to the communications and initiatives they had been exposed to. During the longitudinal, ethnographic immersion in the firm, we conducted unstructured, in-depth interviews with the senior executive as well as interviews with a census of the senior partners, directors, managers, and project leaders. Collectively, this group was responsible for introducing its priority explore and exploit initiatives. Interviews were repeated at three periods during the three-year study, the first at the outset and the second and third during each of the subsequent years. In addition to the Asia-Pacific respondents, interviews were conducted with purposefully sampled representatives from the London, New York, and San Francisco offices in order to assist in establishing an international context. In line with the view that sensemaking needs 'to be complemented by a better understanding of sensegiving' (Fiss and Zajac 2006, p. 1173), we initially focused on leader sensegiving and sensemaking to better understand the language used to frame and communicate the explore—exploit agenda.

In total, more than 50 interview respondents formally participated in the study, offering verbal accounts of their experiences at the outset of the study, after 12–16 months, and after 32–36 months. In addition to the formal interviews, informal conversations and discussions supplemented the data collection process, occurring during the three data collection periods. All formal interviews were video-recorded as well as audio-recorded, the former used by the researchers to assist in reviewing memos and theoretical notes. Handwritten notes were taken summarizing the details of other informal conversations including those undertaken with individuals who voiced their experiences but were not part of the formal interview process. All audio data were transcribed word for word and organized in files held within the qualitative software NVivo 10. This computer-aided software provided structural and practical assistance in the coding of more than 350,000 words of data.

Data analysis involved coding the series of narratives and ethnographic transcriptions into emergent themes reflecting the commentaries of the respondents (Crossley 2007; Riessman 2008). This represented a form of categorical coding analogous to the process of establishing first-order concepts popularized by Gioia and Thomas (1996). Data analysis occurred during the research process and took place after each interview in order to make constant comparisons. Transcripts and video were first broadly studied to gain a general familiarity of the contents. During this process, dominant themes were noted against which each transcript was interpreted and meanings developed. We used the themes to structure and organize the data and analytical process, including the revision of notes, video, and audiotapes.

Data credibility was confirmed through investigator triangulation where two researchers were involved in the interviewing process. As Lincoln and Guba (1985, p. 307) observed, multiple investigators keep each other 'honest.' This process was further bolstered by peer debriefing where the interviewers subjected their working observations to the scrutiny and analysis of a team member not involved in the interviews. In addition to the ethnographic data, contextual resources such as annual reports, written descriptions of the physical environment, historical developments, and current issues identified on the firm's Web site and social media platforms were collected in order to help locate the data within a practical, situational framework.

RESULTS AND DISCUSSION

Sensemaking/Sensegiving Themes

The coding process produced four dominant sensemaking/sensegiving themes, labeled: (1) shaping a dual organizational identity—embracing ambiguity and uncertainty; (2) enabling explore—exploit (duality) thinking—a dual sensegiving/sensemaking endeavor; (3) using sensegiving/sensemaking to engender simultaneous exploration and exploitation; and (4) developing ambidexterity capability—enabling the dual organization. To investigate the measures adopted by the firm to build dual explore—exploit capabilities, we drew on these four themes to underpin our examination of the sensemaking and sensegiving communication strategies of key leaders and implementation managers as they undertook a disruptive change process committed to pursuing exploration and exploitation. The following discussion charts the sensegiving/sensemaking communications trajectory against the four themes.

Shaping a Dual Organizational Identity—Embracing Ambiguity and Uncertainty

By adopting a new communication strategy in which explore—exploit became 'part and parcel of the vocabulary,' leaders at the firm deliberately sought to shape a 'dual' organizational identity. In so doing they aimed to give meaning to new thinking that directly challenged the organization's dominant but conservative, risk-averse approach. The power of managerial language was well appreciated by the CEO: 'An object or a sentence or an expression in people's minds, it becomes all powerful; we have managed over eight and a half years of transformation using simple words that mean something.'

The CEO and his executive at first encountered significant obstacles. After all, they were challenging predictable, entrenched attitudes and behaviors reflecting a fear of change. When novel or untried approaches to a problem were put forward, they were greeted with scorn and negativity. During a series of workshops with senior partners, the CEO decided to confront the negative behaviors head on. He commented, 'I said to them, you know, I'm just sick and tired, these are the sayings that are driving me spare, they're stopping every conversation.' By drawing up a list of sayings that certain workshop participants could be predicted to utter, and presenting these to the senior management group, the CEO shone the spotlight on negative behaviors that 'were stopping every conversation.' This direct, explicit challenge forced a 'moment of truth' and understanding among the senior management group that 'something's got to happen here.'

The success of this new mindset was challenged, however, by the collision of the leadership message with reality. As it became clear to the CEO and his CSO that their message was not resonating with those members expected to implement the changes, the leaders modified their approach. However, they did not attempt to compromise the pursuit of innovation (exploration) in favor of the status quo. Rather, the CEO's message shifted from dealing with explore and exploit ventures as exclusive, highly differentiated organizing elements, to referencing them as entwined, irreducible, and simultaneously essential to survival and prosperity. This represented a shift that has been observed in previous studies (Orlikowski 2007).

For the CEO, being *different*, doing things differently, and working together, were critical to building a more entrepreneurial, innovation mindset. These concepts became central themes: '...we needed a new language, we needed something new... So with the innovation then came how ... what happens if we look at things very differently.' As this comment illustrates, the CEO and his senior executives were keenly aware that the language they adopted was pivotal to selling innovation as a desirable new direction that would allow them to offer a range of novel products. The CSO noted, however, that implementing organizational change is problematic at the best of times: 'When you put forward a new idea, the winner is normally the guy that can kill it the fastest.'

At the same time, the CEO recognized that it was not only the senior partners he had to convince about the need to think and behave differently as an organization. The firm's board also needed to make sense of and be open to new ideas, question conventional thinking, and engage with the new language around innovation and change. The CEO subsequently organized a series of 'yellow hat' meetings (based on the de Bono training principles) to look at new ideas and their 'upside,' considering positive options only. This initiative proved its success as, '20% of our business was born out of that experience.'

Robinson (1981, p. 60) claimed that, 'telling stories about remarkable experiences' is one of the ways in which people try to make the unexpected expectable and therefore manageable. The yellow hat story reveals how the CEO employed evocative language to surface and then challenge the received wisdom about 'the way we do things around here.' Boden (1994, p. 51) similarly depicted organizations as a 'lamination of conversations' where 'decisions are talked into being in fine yet layered strips of interaction.' The yellow hat moment exhorted board-level and senior executive skeptics to be open to new ideas, look for positive solutions, and to encourage organizational members to 'recognize, question and replace' existing practices and behaviors (Boreham and Morgan 2004, p. 309).

Enabling Explore—Exploit (Duality) Thinking—A Dual Sensegiving/Sensemaking Endeavor

Our data suggest that midway through the change process the senior leaders of the firm had begun to accept the necessity of dual explore—exploit thinking. For them, it compelled greater sensitivity and receptiveness to the complexities, ambiguities, and contradictions intertwined in the day-to-day routines of organizational life. By accepting that both explorative and exploitative forms of organizing were essential for success, the senior leaders took the first steps toward enacting a dualities framework and building ambidexterity capabilities.

While the firm's leadership had taken an essential leap, those responsible for implementing the activities had yet to make sense of this new way of thinking that required them to pursue innovation *and* existing efficiencies with equal determination. For example, one director responsible for both delivering upon existing revenue targets and developing new products commented: '…the point of tension between innovation and revenue now… you know it's messy and it's ugly and it's frankly stressful.' In addition, internal research showed that those responsible for implementation did not want to be told *how* to innovate, they wanted to be told *why* to innovate. For one director, being told how to innovate constituted an offense that undermined his sense of belonging: '…why would you go and work for an organization that inherently as a value proposition doesn't trust people?' According to the CEO, '…we then spent the next four months on the why, and silent on the how, and everybody went now we get it and then innovation started happening at a different level.'

In their study of SMEs, Lubatkin et al. (2006, p. 651) found that behavioral integration, through better communication and collaborative activities, engendered trust and reciprocity. This was seen as essential in managing the 'contradictory knowledge processes' essential to achieving a more entrepreneurial mindset unimpeded by the sequencing or the separation of explore and exploit activities. As the CEO observed: '...for me the innovation program and the success of it in getting into the DNA (of the organization) is just how we can inspire people to want to be part of it.'

Sensitive to member feedback and understanding the importance of building trust and open communication channels, senior leaders modified their language from a telling, how-based communication style to a why-driven one that sought to explain more clearly the reasons for change. However, while the CEO recognized the value of stepping back and allowing time for member sensemaking and feedback, he and the senior leadership team were firmly in control of the language employed to shape the change process, signify what was important, and influence meaning making. Their fundamental aim was to make explore—exploit vocabulary and operations 'part and parcel' of the organization, embedding it 'into the DNA.'

For those middle managers, directors, and partners trying to make sense of the messages from the CEO and CSO, the shift struck a chord. For example, one director began with concerns over the skills of her staff and the available resources: 'Many of them don't have the skills to deliver or certainly execute on all of those skills.' However, the 'why' approach to innovation changed her perspective. Innovation itself became the solution to the problem: '...so we'll need to bring in some more resources.'

Using Sensegiving/Sensemaking to Realize Simultaneous Exploration and Exploitation

While the firm set about tackling the complexities and challenges of fostering innovation, the senior leadership also recognized that it was important 'not to slip' on initiatives within the traditional exploit areas of business. The CSO commented, '...we had to exploit a lot to deliver the profits that we promised our partners, but then how do we explore and how do you evolve for the future? We got to the point of both explore—exploit becoming part and parcel of our vocabulary and ... of the way we started to think about things.'

Initially, the explore and exploit arms of the business were treated as separate operating systems and structures. But, over time a belief emerged that greater benefits were to be realized by somehow bringing the two together. One senior partner recognized that integration needs 'gates' rather than 'silos.' There must be enough 'glue' between explorative and exploitative areas of the business to ensure that when an innovative idea becomes 'commercially relevant,' the technical experts on the exploit side of the business can test it as quickly as possible. The senior partner emphasized an exploit mindset 'over all innovations from very early on.' Otherwise the danger is 'they just stay out there' and their potential is never realized.

As another senior partner argued, the traditional input and support from the exploit side of the business is critical if innovation is to flourish. The organization learned that without a 'vibrant, successful' exploit platform to work from, innovation and experimentation are not possible: '…bringing in new thinking, new streams of revenue, changing the way we deliver the old, okay, but that doesn't mean that the core business disappears.'

Reflecting a dualities mode of thinking, the preceding examples demonstrate that innovation had started to be viewed as complementary to existing businesses rather than contradictory. In order to grow and 'be different,' the firm had to excel at *exploring* new opportunities as well as *exploiting* existing products and services. According to one senior partner, as an innovation project gained momentum and moved to commercialization, 'chameleon like, the leader [innovation champion] has to transform.' The challenge now was to harness the growing commitment to pursue exploration and exploitation and convert the ideas from aspirations to operations. This necessitated building ambidexterity capabilities. In the next section, building capacity for the equal, skilled pursuit of exploration and exploitation is considered against four factors that appeared significant. These four factors reflect the measures outlined in Chap. 3 for developing ambidexterity capability.

Developing Ambidexterity Capability: Enabling the Dual Organization

Resourcing and Supporting Business Unit Leaders

The earlier description of the business unit leader as a 'chameleon,' charged with a transformation, underlines the criticality of matching sensegiving communication with sensemaking interpretation and feedback. Business unit leaders had to become willing and confident participants in the change process. Managers play a central role in creating linkages to communicate a shared vision (Gibson and Birkinshaw 2004; Lubatkin et al. 2006; Turner et al. 2012). Therefore, as business unit leaders in turn become sensegivers, they too must be equipped with the knowledge and communicative ability to inform and explain new initiatives to business unit members.

A cultural change transpired within the firm recognizing the importance of both leader sensegiving and member sensemaking to ensure a shared change direction. An initial skeptic, one director suggested that they were '…creating a new culture… that frees people to think about things differently and to also make sure that actually we step out of telling people what we know.' It was clear to the senior executive that partners accustomed to leading traditional exploit-based business units not only needed to be equipped with appropriate resources, skills, and training in order to understand the issues around building dual ambidextrous capability, but they also needed to be comfortable working on a daily basis with a level of ambiguity. There was a widely held view that organizational members needed to become more agile and flexible. As one partner put it, they had to 'be prepared and happy to work in the grey.' Traditional ways of thinking tended to impede experimentation and risk-taking: 'This role of the entrepreneur is a really difficult one because … you need to behave in a much more agile way than the core business does.' Organizational members were urged to 'turn down the noise of their technical expertise' and appreciate that uncertainty, risk-taking, and the unknown accompany both innovation and efficiency.

Effective Leader Communication as Explore—Exploit Advocates
The idea of pursuing exploration and exploitation as dual, interdependent endeavors became a central part of the sensegiving/sensemaking language at the firm. In one comment illustrating the change effect, one director described the duality-infused communication strategy as '…really helpful as a frame for our leaders to do that. And it's changed the belonging.'

Alongside the increasing stakeholder engagement with the innovation program and growing understanding of *why* they needed to think and work differently, pockets of dissent still existed among a number of the traditional areas of the business. A senior partner and innovation champion referred to two 'factions' in one core business that paid lip service to the program, and in fact, 'fought against their own.' There was a 'bifurcation of effort' with one group arguing that they were operating in an *exploit* strategic workforce and the other group arguing that they were operating in an *explore* strategic workforce. A senior partner noted: 'So we learned very early on that … to actually make it work you have to have … an ambidexterity.'

While overall there was a significant commitment made to the dual explore—exploit endeavor, organizational members needed educating on how to operationalize both simultaneously, and with equal skill and dexterity. As one senior partner explained, business unit leaders and organizational members needed to be able to think through issues from two different perspectives: 'We need to be able to dial up and dial down … you have to be able to flick … switch between perspectives.'

Human Resources Initiatives to Develop Socio-technical Skills and Confidence in Managing Explore—Exploit Tensions

Adjusting to the new thinking proved challenging for partners accustomed to leading traditional exploit-based business units. They needed to be equipped with appropriate resources, skills, and training in order to understand the issues around operating in explore and exploit sides of the business at the same time. As the senior partners' observations earlier indicate, they needed to develop confidence and comfort with an ambiguous set of organizing structures. As previous research has shown, the key requirements for dealing with complex, uncertain environments include access to effective top-down/bottom-up and lateral knowledge flows, a capacity to analyze and interpret both short- and long-term issues, and a robust prior-related knowledge and experience base that fuels curiosity and facilitates new knowledge (Raisch et al. 2009). Moreover, as Orlikowski (2007, p. 1438) argued, the right mindset needs to encompass both social (human-centered) and material (technological) factors as they 'relationally entail or enact each other in practice.' Dual attributes need to be factored into mentoring and leadership training programs that work outward to reach socio-technical work practices. As identified in Chap. 3, while structural factors were important, contextual/behavioral factors proved essential to achieving ambidextrous capability.

Development of Appropriate Performance Measures

Business unit leaders must be equipped with the knowledge and communicative (sensegiving) ability to inform and explain new initiatives to business unit members (sensemaking), not least how they will be measured and rewarded. As Jansen et al. (2008, p. 999) proposed, achieving organizational ambidexterity demands not only 'a strong and compelling shared vision,' but also a 'shared fate' in the form of contingency rewards to ensure a combined explore—exploit approach. Developing dual performance measurement must be tackled early to ensure organization-wide acceptance and commitment to both exploration and exploitation.

As the case experience showcases, appropriate performance and reward measures that take into account the distinctive facets of exploit and explore activities need to be identified, explained, *and* implemented. For the firm, the new reward structures became a core component in the sensemaking process. Building the capacity to pursue explorative opportunities while adapting and enhancing exploitative initiatives with equal enthusiasm remains an ongoing challenge for the firm. As the CEO observed 'you only have explore if you survive exploit.'

THEORY TO PRACTICE

An explore—exploit organizing form that pivots on 'stability-maintaining' change demands a dual sensemaking and sensegiving communication strategy. This sets the context for ambiguity, uncertainty, and contradiction, in which seemingly stable organizational routines are not immune from disruption and regeneration (Brown et al. 2009; Tsoukas and Chia 2002).

The CEO recognized that in order to give meaning to the duality message, he and his senior executive needed to pitch a convincing story that embodied a radically different business model. Moreover, the message delivered by the senior leadership had to adapt to the responses of sensemakers charged with implementing the new order. The new vision gained traction as sensegiving and sensemaking were both recognized as important and integral, impelling new ways of thinking consistent with ambiguity and uncertainty.

Over time, as the sensegiving/sensemaking communications strategy evolved, the new ways of thinking were reinforced by tangible structures and systems including rewards for both explore and exploit success. However, although change transpired as sensegiving and sensemaking were enacted as mutually reinforcing communication strategies, the theoretical mechanisms guiding the successful transition from duality theory to the practice of simultaneous modes of exploration and exploitation remain more elusive, or at least, more difficult to generalize.

The work of Sutherland and Smith (2011) highlighted some possible explanations for the absence of a concrete theory explicating the mechanisms and variables governing the simultaneous management of innovation and control. They suggested that part of the obstacle lies in the dominant perspective of large organizations. Rigid bureaucratic organizing forms are the default, but tend to introduce more flexible forms at the same time. The problem with this orientation is that innovation is being mandated by top-down interventions, when in practice the evidence indicates that the most successful innovation is characterized by emergence (Graetz and Smith 2008). The catch is that emergence requires the right environment, the removal of boundaries, and the reduction of risk—steps that emanate from the top-down. We think this helps explain why the combination of a duality theoretical lens and a narrative sensegiving/sensemaking interpretive design provides such rich insights into the explore—exploit problem.

CONCLUSION

In proposing a new duality-driven organizing reality where exploitation and exploration operate in active tension, the case revealed that leader sensegiving must offer opportunity in the midst of flux and uncertainty. It must also be delivered in a language that *makes sense* to its audience. Importantly, the leader sensegiving/sensemaking communications strategy needs to shift from one formulated around constancy, efficiency, and control to one imbued with duality thinking (Graetz and Smith 2008). Such an approach seeks to preserve organizational identity, and a sense of order and continuity that simultaneously encourages frame-breaking initiatives promoting innovation and change. It demands the capacity to *explore* for new knowledge while simultaneously *exploiting* existing knowledge (March 1991).

As Bruner (1991) maintained, organizational communications concern people acting in a setting. Consequently, it is important that leaders, through their sensegiving endeavors, ensure that the 'happenings' that organizational actors experience parallel their 'intentional states'—their beliefs, desires, attitudes, and values (Bruner 1991, p. 7). This case reveals that the power of leader sensegiving resides in its ability to persuade, influence, cajole, and shape member sensemaking through the actions, words, ideas, and images they communicate with. As one senior partner observed, it is the 'restlessness' of the leadership group to actually make something happen and to challenge the status quo that shaped the new language of exploration and exploitation.

Our data support the premise that organizational change can be canalized by symbolic communication strategies that guide meaning making. From this perspective, studying sensegiving and sensemaking offers an insight into the effects of formal and informal structures as seen through the covert inner workings of individuals engaged in the process of constructing meaning. Organizational members do not engage with change in a vacuum. Language communication plays an essential sensemaking function for organizational members attempting to find their way in a complex environment infused with ambiguity and uncertainty. Our case study has revealed how one firm successfully introduced ambidexterity capabilities by using a method of communication consistent with, and made transparent by, sensegiving and sensemaking.

In addition, based on the developmental initiatives taken within the firm, we outlined what emerged as initial steps toward building ambidexterity capability to deliver dual explore—exploit outcomes. These included: effective leader communication as explore—exploit advocates; establishing appropriate resources and incentives for business unit leaders to support innovation and control systems and structures; putting in place targeted human resources initiatives; and introducing appropriate performance measures that recognize and reward both explore and exploit initiatives. In the following chapter, we examine another case where an organization devised a 'fluid' approach to building ambidextrous capabilities and achieving a dual organization.

REFERENCES

Berger, P. L., & Luckmann, T. (1967). *The social construction of reality: A treatise in the sociology of knowledge.* New York: Anchor Books.

Birkinshaw, J., & Gibson, C. (2004). Building ambidexterity into an organization. *Sloan Management Review, 45*(4), 47–55.

Boden, D. (1994). *The business of talk: Organizations in action.* Cambridge, MA: Polity Press.

Boreham, N., & Morgan, C. (2004). A social cultural analysis of organizational learning. *Oxford Review of Education, 30*(3), 307–325.

Brown, A. D., Gabriel, Y., & Gherardi, S. (2009). Storytelling and change: An unfolding story. *Organization, 16*(3), 323–333.

Bruner, J. (1991). The narrative construction of reality. *Critical Inquiry, 18,* 1–21.

Crossley, M. (2007). Narrative analysis. In E. Lyons & A. Coyle (Eds.), *Analysing qualitative data in psychology* (pp. 131–144). Los Angeles: Sage.

Denis, J.-L., Lamothe, L., & Langley, A. (2001). The dynamics of collective leadership and strategic change in pluralistic organizations. *Academy of Management Journal, 44*(4), 809–837.

Dunford, R., & Jones, D. (2000). Narrative in strategic change. *Human Relations, 53*(9), 1207–1226.

Farquhar, J. D. (2012). *Case study research for business.* Thousand Oaks, CA: Sage.

Fiss, P. C., & Zajac, E. J. (2006). The symbolic management of strategic change: Sensegiving via framing and decoupling. *Academy of Management Journal, 49*(6), 1173–1193.

Fleming, D. (2001). Narrative leadership: Using the power of stories. *Strategy & Leadership, 29*(4), 34–36.

Gibson, C. B., & Birkinshaw, J. (2004). The antecedents, consequences, and mediating role of organizational ambidexterity. *Academy of Management Journal, 47*(2), 209–226.

Gillham, B. (2000). *Case study research methods.* London: Continuum.

Gioia, D. A., & Chittipeddi, K. (1991). Sensemaking and sensegiving in strategic change initiation. *Strategic Management Journal, 12*(6), 433–448.

Gioia, D. A., & Thomas, J. B. (1996). Identity, image, and issue interpretation: Sensemaking during strategic change in academia. *Administrative Science Quarterly, 41*(3), 370–403.

Graetz, F., & Smith, A. (2008). The role of dualities in arbitrating continuity and change in forms of organizing. *International Journal of Management Reviews, 10*(4), 265–280.

Hill, R. C., & Levenhagen, M. (1995). Metaphors and mental models: Sensemaking and sensegiving in innovative and entrepreneurial activities. *Journal of Management, 21*(6), 1057–1074.

Jansen, J. J. P., George, G., Van den Bosch, F. A. J., & Volberda, H. W. (2008). Senior team attributes and organizational ambidexterity: The moderating role of transformational leadership. *Journal of Management Studies, 45*(5), 982–1007.

Jennings, G. (2001). *Tourism research.* Milton, Australia: Wiley.

Lincoln, Y. S., & Guba, E. G. (1985). *Naturalistic inquiry.* Beverly Hills, CA: Sage.

Lubatkin, M. H., Simsek, Z., Ling, Y., & Veiga, J. F. (2006). Ambidexterity and performance in small-to medium-sized firms: The pivotal role of top management team behavioral integration. *Journal of Management, 32*(5), 646–672.

Maitlis, S., & Lawrence, T. B. (2007). Triggers and enablers of sensegiving in organizations. *Academy of Management Journal, 50*(1), 57–84.

Orlikowski, W. J. (2007). Sociomaterial practices: Exploring technology at work. *Organization Studies, 28*(9), 1435–1448.

Peirano-Vejo, M. E., & Stablein, R. E. (2009). Constituting change and stability: Sense-making stories in a farming organization. *Organization, 16*(3), 443–462.

Raisch, S., Birkinshaw, J., Probst, G., & Tushman, M. L. (2009). Organizational ambidexterity: Balancing exploitation and exploration for sustained performance. *Organization Science, 20*(4), 685–695.

Riessman, C. (2008). *Narrative methods for the social sciences.* Los Angeles: Sage.

Robinson, J. A. (1981). Personal narratives reconsidered. *Journal of American Folklore, 94*(371), 58–85.

Simsek, Z., Veiga, J. F., Lubatkin, M. H., & Dino, R. N. (2005). Modeling the multilevel determinants of top management team behavioral integration. *Academy of Management Journal, 48*(1), 69–84.

Smith, W. K., & Lewis, M. (2011). Toward a theory of paradox: A dynamic equilibrium model of organizing. *Academy of Management Review, 36*(2), 381–403.

Sonenshein, S. (2010). We're changing—Or are we? Untangling the role of progressive, regressive, and stability narratives during strategic change implementation. *Academy of Management Journal, 53*(3), 477–512.

Strauss, A., & Corbin, J. (1990). *Basics of qualitative research: Grounded theory procedures and techniques.* Newbury Park, CA: Sage.

Sutherland, F., & Smith, A. (2011). Duality theory and the management of the change-stability paradox. *Journal of Management and Organization, 17*(4), 534–547.

Tsoukas, H., & Chia, R. (2002). On organizational becoming: Rethinking organizational change. *Organization Science, 13*(5), 567–582.

Turner, N., Swart, J., & Maylor, H. (2012). Mechanisms for managing ambidexterity: A review and research agenda. *International Journal of Management Reviews, 15*(3), 317–332.

Weick, K. E. (1995). *Sensemaking in organizations.* London: Sage.

Weick, K. E., Sutcliffe, K. M., & Obstfeld, D. (2005). Organizing and the process of sensemaking. *Organization Science, 16*(4), 409–421.

Yin, R. K. (2009). *Case study research—Design and methods* (4th ed.). Thousand Oaks, CA: Sage.

CHAPTER 5

Structuring Innovation

Abstract This chapter introduces a second case study exposing a complex network of simultaneous explore and exploit activities undertaken by a large firm over a decade. It depicts a response to the explore—exploit paradox where switching emphasis and resources between the two priorities failed, leading to a novel combination of heavy exploitation-driven actions alongside deep exploration projects. The chapter examines how the firm's success emanated from its fluid organizing forms approach to dealing with the explore—exploit tension. Instead of seeking to constrain the tension, the firm escalated it into a productive and dynamic source of innovation. Fluidity, the chapter concludes, commands a central place in fostering an ambidexterity-friendly environment

Keywords Fluid innovation · Explore—exploit tension · Case study

INTRODUCTION

This second case-driven chapter reveals a complex network of simultaneous explore and exploit activities undertaken by a large firm over a decade. It depicts a response to the explore—exploit paradox where switching emphasis and resources between the two priorities failed, leading to a novel combination of heavy exploitation-driven actions alongside deep exploration projects. Most pointedly, the organizing response was fluid. Our interpretation suggests that a large part of the case firm's success lay with their fluid organizing forms approach to dealing with the explore—exploit

© The Author(s) 2017   75
A.C.T. Smith et al., *Reinventing Innovation*,
DOI 10.1007/978-3-319-57213-0_5

tension. Instead of seeking to constrain it, they sought its escalation into a productive tension, powerful enough to impel individuals to innovate, but also sufficiently contained to be captured at an organizational level. As we noted in earlier chapters, according to the literature, a core challenge in reconciling the tension inherent in managing innovation and commercialization structures pivots around a misalignment between the organization and the individual; there has been an overreliance on developing the organizational mechanisms needed to enable ambidexterity capabilities with little appreciation or understanding for the importance of individual fluidity. Although it might work as an abstract, even theoretical concept, in practice individuals tend to find it difficult to excel at both exploitation and exploration. As a result, they must manage contradictions and conflicting goals, work with uncertainty and ambiguity, be comfortable taking risks, and perform diverse, swiftly changing roles. This chapter highlights a series of organizing forms in which individuals can find it easier to be both innovative and conservative. In short, we chart the initiatives of an organization as they created an environment conducive to developing ambidexterity capabilities.

Like most large organizations, our case firm needed to find a rare combination of groundbreaking innovation while maintaining a successful operational performance, or what we have referred to as the explore—exploit conundrum. Employing 'grounded research,' we collected data on the experiences of the firm's key decision makers, innovators, and 'intrapreneurs.' Our findings revealed a three-tiered organizing forms response to the explore—exploit paradox, leading us to conclude that one successful approach focuses on the productive tension that emerges by enacting innovative organizing forms within conservative boundaries. We also outline an explore—exploit framework that conceptualizes organizational changes incorporating complexity and contradiction, without the implicit emphasis on removing or denying the existing tension.

To foreshadow, our results reveal a complex network of concomitant explore and exploit activities. Most pointedly, the organizing response was fluid. For example, the firm created a new major business unit, liberated it from the demands of normal operations, placed an entrepreneurial maverick in charge, and supported it like a start-up. As the unit became successful, it was spun back into the mainstream business, and another, more advanced iteration spun out again as another change stimuli. Not only was the approach successful in driving new innovation, but it also had a powerful cultural impact on the normally intractable exploit side of the business. The deeply conservative business clusters driven by incremental monthly targets started to engage with the firm's 'rise and fold' innovation projects. Fluidity, we conclude, commands a central place in fostering an ambidexterity-friendly environment.

In the following sections, we profile the study's research design before presenting and discussing the results of our analysis. The chapter culminates in a model summarizing the case firm's approach to building a dual organization, leading to sections offering theoretical and practical implications, followed by conclusions.

Case and Context

This chapter took its initial motivation from the tangible success the case firm produced over a 10-year period—encompassing two global financial crises—culminating in its emergence as a billion dollar firm with an envious growth rate. This case chapter asks how a large firm delivers on efficiency and control at the same time as flexibility and innovation through its organizing forms. To answer this question, we began with an interpretive mindset (Isabella 1990), assuming that organizational actors create and enact their own social realities (Berger and Luckmann 1967). Furthermore, their realities are likely to be shared and heavily influenced by senior members as well as by the maturing, meaning-making nature of time and reflection (Deetz 1996). Context was therefore sovereign, our approach mirroring Galvin's (2014) description of an 'inside-out' research focus where the chief issues hold particular salience within a specific environment. In building a model to describe the case organization's response, our aims go beyond understanding the single context as the results provide insight for all organizations.

Case studies enable the identification of dynamics present within single settings (Eisenhardt 1989), operate under specific contexts (Daymon and Holloway 2002), and require detailed, rich data in order to be understood. The study utilized data collected from in-depth interviews, supplemented by data obtained from ethnographic techniques over a period of 24 months. Data were analyzed with grounded methods consistent with Whiteley's (2004) 'grounded research.' It departs from a steadfast grounded theory methodology in the Glaser and Strauss (1967) conceptualization, in favor of a 'pragmatic' (p. 32), modified version attuned to the business setting, but holding faithful to 'theoretical sensitivity' where emergent data are considered in light of existing theory. As a result, the naturalistic inquiry approach utilized in this research employed a conceptual coding method (Strauss and Corbin 1990). Our approach was guided by Denis et al. (2001), who like Whiteley (2004) outlined how nascent theory can be explored and developed by starting with a theory-inspired problem and pursuing it with a detailed inductive procedure. We began with the explore—exploit problem inspired by duality theory and used conceptual coding to compile the data into themes relevant to the case firm's response to creating ambidexterity capabilities.

As an inductive procedure, grounded research seeks to identify conceptual categories arising from the data using what Glaser and Strauss (1967) named the constant comparative method. It works through an ongoing interrogation of the incidents observed in the data and the emerging theoretical concepts. We compared the contents of each of the interview's data units as well as 24 months of ethnographic observations as they were examined, with the aim of specifying the common and diverging themes. Our procedure employed theoretical sampling to pre-select each forthcoming interview, adding depth and perspective to the emerging codes. As themes evolved and matured, common concepts became apparent, which were subsequently used to generate data categories. The categories provided the foundations for theory development, thereby finalizing the process of theoretical sensitivity. Thus, as our codes demonstrate, the focus was upon capturing and explaining the action and interaction strategies of organizational actors (Strauss and Corbin 1990).

Unstructured, in-depth interviews were conducted with the senior executive of the Asia-Pacific arm of the firm, as well as interviews with a theoretical sample of senior partners, innovation directors and managers, and project leaders across the firm who played pivotal roles in shaping, and responding to, its organizing form initiatives. One specific business unit of the firm, a unique, semi-autonomous innovation-driven entity, was examined through an interview census of its members. In addition, ethnographic data were collected through a series of participant observations undertaken over several months. Finally, interviews with purposefully sampled representatives liaising with the firm from several international offices were also undertaken to assist in establishing context. In total, 32 formal interviews were conducted, transcribed and coded, excluding the numerous repeat interviews informally undertaken as part of the participant observation. All audio data were transcribed and organized within the qualitative software NVivo 10. By constantly comparing data, a model evolved, subject to interrogation at successive iterations of data collection.

The model focuses on relationships between explanatory variables, seeking to establish the causal antecedents; it aims to predict and explain how one company designed an environment for ambidexterity, in the process creating a dual organization.

RESULTS AND DISCUSSION

We present the data by firstly identifying and describing the four major (open) codes and their general contents. The first open code labeled *Models and Concepts* captures the paradigms, programs, or constructs employed by the firm. These models and concepts represent shared mental lenses depicting key assumptions about the firm's internal engagement and ways of perceiving the outside world. In addition to views about its culture and values, respondents expressed a keen awareness of the need to shift away from traditional organizing forms and work with models that embrace the tension associated with the explore—exploit imperative. As a consulting partner and member of the firm's Innovation Board observed, 'Is there sufficient uncertainty around it? If there's not enough uncertainty, it probably should be done again, another model.'

In turn, eight second-level axial codes conceptualized *Models and Concepts*: (1) concept stage; (2) organizational culture; (3) programs and products; (4) idea-ideation; (5) H1, H2, H3; (6) innovation; (7) explore—exploit; and (8) learning. One axial code requiring further comment is 'H1, H2, H3.' This refers to the three-horizon model: The first horizon (H1) involves minor, incremental innovations designed to bolster current operations; horizon two (H2) innovations focus on extending competencies into new markets or commercializing nascent, high-potential new products; and horizon three (H3) deals with innovations capable of radically changing the composition of a firm's product/market position. However, as the CEO observed in framing why the firm was forced to experiment with untested organizing forms, conceptual models do not easily translate into the real world: '...it was supposed to be the perfect model of H3 into H2 into H1; we very quickly saw that this thing doesn't really work in H2. Let's put it aside and run it very, very separately.'

The second open code, labeled *Actions*, was the largest and represented the translation of the firm's 'models and concepts' to its operational, organizing forms focus from 2003 until 2013. One senior partner commented: '...what was recognized three years ago was that there was a problem that was unanswered and a willingness to invest and explore that.'

Actions comprised two axial codes: (1) performance and delivery and (2) marketing. Performance and Delivery contains 10 third-level selective codes and eight fourth-level selective subnodes, mainly reflecting the firm's perceived and actual success in delivering on its new fluid structures. According to the firm's Innovation Director, '...we've got the strategy, we've said we're going to do what we're going to do at a high level, we're now going to take it down another level in terms of the initiatives and building project plans.' Marketing contains 13 selective codes and 10 selective subnodes, principally dealing with the firm's aggressive brand position, operational processes, and leadership underpinning the choice of organizing forms, or what the Chief Financial Officer (CFO) described as '...big plays that we thought could make a difference to our growth and branding.' That is: 'Being known as the most innovative firm is really important' (Senior Partner).

The third open code, labeled *Interactions*, concerned the firm's approach to engagement with internal and external stakeholders. Five axial codes comprised *Interactions*: (1) client Interface; (2) executive level; (3) external interface; (4) staff level; and (5) tension. Throughout all forms of engagement with internal and external individuals and groups, the tension arising from the dual need for explore and exploit activities proved a common element in respondents' comments. With the appetite for change from the leadership came the mandate to experiment and challenge conservative ways of organizing, including the widely held assumption that new structures should at least be harmonious and permanent, rather than deliberately contradictory and transient. Allied to this desire for change was a kind of certainty that if the firm could find the right organizing drivers for both explore and exploit, they would reap the rewards. According to the CEO, '...we have to change, we are going to gain ground like we've never seen before if we now act out the agility that we have been training for since 2003.' However, with new initiatives and expectations also came pain. For one senior partner at the coalface of a small, new start-up venture project, only part of the equation seemed successful. She commented, 'Well we are failing cheaply, like it's costing nothing for me to spend all my extra time ... I'm slowly getting burnt out.'

The fourth open code was *References*, capturing sources of influence on the firm's path from 2003. It was clear that importing the explore—exploit and horizons concepts had caused a significant impact, infiltrating the thinking and everyday language of the respondents, including the Chief Strategy Officer, Asia-Pacific: 'The three horizon model ... which is saying in H1 that's where exploit happens and in H3 that's where explore happens and you have to have a healthy pipeline.'

References included three axial codes: (1) commercialese; (2) management thinkers; and (3) success and failed organizations. Collectively, these three themes reflected the new language and mindset that had infiltrated the company. One senior executive noted: 'We started to take some of the lessons again from the latest work … which is basically saying, you know, think big but bet small and the only way it can get together is to be disruptive, and that is what you should do in H3 and it should not cost a lot of money.' The firm's leadership had spent considerable time and money traveling to prominent management schools to acquire the latest tools and thinking: 'I then took the executive all to Harvard a year later to a course on managing organizational change and renewal, which is really about the theory that you can take innovation into an existing company' (CEO). At the same time, the CEO and his executive team remained vigilant about lessons from both failures and successes: 'Kodak is my great inspiration for not wanting to be like them.'

An Explore—Exploit Model

The model reproduced in Fig. 5.1 conceptualizes the results and explains the unique organizing forms model that the firm evolved to secure both explore and exploit outcomes. The first part involves aggressive initiatives in both explore and exploit spaces simultaneously. The second involves the implementation of horizon two projects designed to fast-track the scaling-up process from innovation to commercialization. The final part involves the creation of a new business unit operated as a start-up and its subsequent return to the mainstream business, after which the process was repeated.

The model explains the firm's response to its explore—exploit challenge. In Fig. 5.1, the horizontal lines at the top and bottom of the page specify the firm's contextual position, the former through its internal modes of thinking and operation, and the latter through the external references influencing judgment and action. Internally, the company operationalized its philosophy through a bold strategic objective: '30% of revenue will come from new or substantially different service offerings every two years.' As a billion dollar company, this meant generating more than $300 million of new business services every two years. As noted in the figure, the ambition was for the company's operations to assume the number two position in the market.

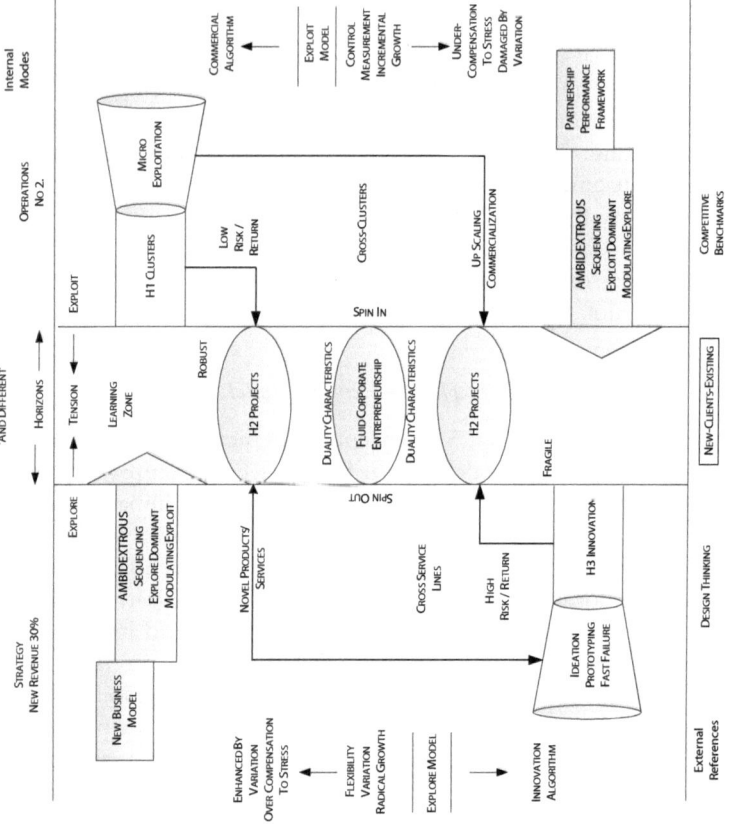

Fig. 5.1 Explore—exploit model

The Chief Strategy Officer reinforced the importance of the strategy: 'It replaces the stuff that becomes redundant and commoditized that we need to exit, but it also creates growth opportunities, which our people need like humans need oxygen.' Simultaneously, the firm distanced itself from the natural inclination to prescribe and inform, '...we were talking about what new services means ... how it creates opportunities for growth for individuals, how many people actually make partners out of that, how this is cool to be in the new space.'

External influences also shaped the burgeoning entrepreneurial approach. While benchmarks provided by competitors and lessons from other successful companies were studied intensely, the firm also imported a 'design thinking' method as a tool toward a renewed mindset for approaching clients, products, and the new organizing forms that were needed. According to the Innovation Director, 'Design thinking is about getting people to think differently, really getting them to look at things and change the way they look at things today.' Even the once skeptical CFO observed that, 'Things we're doing around design thinking now are starting to resonate.'

Theme 1: Fluid Dualities and Innovation/Commercial Algorithms

In recognizing the problems associated with oscillating attention and resources between explore and exploit, the firm crafted a new organizing framework where the two modes of structuring would be pursued. But, rather than a balance, high levels of both were sought at the same time, which also led to a more fluid approach to organizing than they were used to. Whereas in the past new structures were put in place on the assumption that further change would be in the distant future, the firm now began with the principle that change to structures would be constant and essential in order to capitalize on the tension between the distinctive organizing forms associated with control and flexibility. According to the Chief Strategy Officer, '...the exploit process said you put a business case in place with certain workflow, this is what you do, this is the approvals you get, get it signed over 10 places.' Conversely, in the explore process '...you come up with a good idea, you get $10,000 to go and play with, no questions asked.'

In the exploit sphere, illustrated on the right of Fig. 5.1, the firm divided its operational business units into 60 highly granularized, specific, measureable and accountable 'clusters.' Performance in each cluster was assessed with forensic vigor: 'You just focus on what we call operational excellence measures' (CFO). The result was the division of the clusters into five performance bands—fix, enhance, optimize, accelerate, and dominate—designed to encourage the elevation of lesser performing groups at the same time as revealing where to invest and grow those of the greatest potential. As one senior partner indicated, for those underperforming clusters, 'We are not interested in your growth, your client penetration, because we think that for every dollar, for every new client you get, we lose a dollar, so fix your basic operating model.' This micro-exploitation represents the firm's commercial algorithm.

On the explore side of the ledger, depicted on the left in Fig. 5.1, the firm invested heavily in organizing forms dedicated to driving innovation. As early as 2004, an internal innovation program was established, designed to provide employees with a forum to share and generate ideas. The program liberated any individual willing to mobilize their ideas into a reality: 'But the beauty of this is you give a person $10,000, they spend their weekends on this. You get $50,000 worth of input for that investment. That was the best return on investment we ever got because we tapped into discretionary effort, and people didn't feel abused. They loved it, and that now is core to the way that we look at our innovation program' (CSO).

Theme 2: Horizon 2 Projects as Scalable Organizing Forms

The center channel of Fig. 5.1 depicts the 'productive tension' between explore and exploit. Within this channel are horizon two (H2) projects, designed initially as commercialization pathways to upscale high-potential innovations as fast as possible and return them to the business clusters. As such, they rely on being transient organizing forms for success: 'We define success in Horizon 2 as the speed with which we can create a scaled business to go back into the core ... so they're fairly attractive, got good scale and there's some uncertainty, and that perhaps means there's more than one service line involved' (H2 project leader and partner). Under this original conception, clusters were considered inappropriate hosts for H2 products, too conservative, static and time-poor to invest in nascent ideas. Equally, the innovation program was too informal and unstructured to accommodate the tight market expectations of clients expecting a rigorous and tested new product: 'H2 requirements and the scale and the complexity, it's got to be more about future trend spotting than coming out of incremental strategic insights ... that's not to say that they wouldn't have input from the innovation program but I think it's a little less likely' (H2 project leader and partner). As a result, the firm needed an organizing form that offered the solidity and confidence of a more formal structure without relinquishing the fluidity and agility of an ad hoc structure.

After several years of difficult experimentation—'…you know it's messy and it's ugly and it's frankly stressful'—the H2 projects began to evolve naturally into two streams, mainly because an independent program was less influential across the firm than semi-independent projects receiving support from a central group but connected more directly to an innovation or a commercial channel. One set of H2 projects reflected a high-risk and return orientation, having grown as a radical new innovation seeking a commercial home. Another set aligned more directly with a sponsoring cluster and sought to infuse a new product or market orientation around an already-successful initiative. According to the CEO, '…it'll be the third year in a row that H2 has not got the support I believe it deserves to get. We're going to persist…' The overarching problem was that the projects required partner champions to shift focus from their profitable cluster businesses ('I gave up a 4.5 million dollar ledger'). As a result, 'the biggest challenge is that you have a very small group of people and skill wise you're loading … there's high expectations loading up onto them' (H2 leader and partner).

The idealized intention of a series of middle-ground H2 projects encountered difficulty, belonging to neither camp, and supported by no one other than a small, centralized support unit. But two developments added traction to the ongoing H2 projects. First, those effectively up-scaling existing products in the form of a significant product/market expansion began to align with a parent cluster, keen to incorporate the addition as soon as possible, fueled by the powerful and ever-present reminder that 30% of their business must be replaced every second year. Second, the more radical innovations, too far from commercial success for a cluster to engage with and typically involving high levels of technical specialization or digital knowledge, began to connect with a relatively new, independent start-up business unit being operated by a brazen internal entrepreneur, conferred with unprecedented autonomy and a propensity to make things happen fast. The cluster leaders had found a way to meet their confronting new targets for growth, while the innovators had found a vehicle to progress their ideas without falling victim to analysis and measurement.

Theme 3: Fluid Organizing Forms for Optimizing Duality Tension

Figure 5.1 presents 'fluid corporate entrepreneurship' at the center of the tension between explore and exploit. This third theme reflects the development of a spin-out. Unlike a typical corporate spin-out it remained within the company but, at the same time, operated like an entreprencurial start-up.

The spin-out was formed in 2009, ostensibly to create digitally driven professional services. The internal start-up assumed a platform emphasizing technology-based services and stand-alone products, but its real differentiation was an entrepreneurial method emphasizing rapid prototyping, proof of concept, fast and cheap failure, and agile systems. At the helm was a partner described by the CSO as someone specializing in 'pushing the boundaries.' This CEO reveled in the prospects of entrepreneurial empowerment, knowing from his own successful business ventures that 'Innovators don't necessarily want rewards, they want to see their idea executed. You give them a bit of oxygen, you give them a bit of capability, you help them get something out the door, they'll put in.'

Aimed to be a 'game changer,' the spin-out challenged the traditional firm model. Instead of high margins and few clients with high touch, regular face-to-face contact, the new model worked on low touch, low resolution, infrequent face-to-face contact, self-service, high volume of transactions, and numerous clients at low margins. Under the revised model, generating new ideas was subservient to their transformation into rapid execution. The small team comprised a mix of young entrepreneurs and more experienced members recruited from traditional clusters.

In 2012, the spin-out blended into the exploit world aiming to consolidate a suite of integrated digital products and services. For example, both product and market opportunities expanded radically when the spin-out's new tools were wielded by other parts of the firm. The organizing model had come full circle: the firm spun out until it had secured its mandate and then spun back in. Innovation needed freedom until it became sustainable but then needed control to generate scope and efficiency. One of the firm's managing partners added: 'We've brought out what we had as our online practice which was our consigned practice that takes external digital offerings to the market, Web design development, e-commerce. And we've brought them both together now.' As a result, a world of new services was opened by the digital bridge: '...we've got a human capital practice over here that consults on leadership, let's bring those two together and digitize and differentiate your leadership offering that you take to market.' With the firm's opportunities for contained but disruptive innovation curtailed by the absorption of the spin-out into main operations, its former CEO was assigned to repeat the process in spearheading a new innovation venture.

Theory and Practice: Fluid Organizing Forms

The firm's model of three organizing forms offers a workable response to the explore—exploit tension. However, rather than an opportunistic, dynamic oscillation between innovation and efficiency, the model reveals how the two can be practiced simultaneously through a form of 'fluid' structures. One initiative saw the creation—and most importantly liberation and return—of a spin out. As a genuine start-up, the business unit broke all the rules, bound not by financial targets, but rather by a commitment to experimentation, fast and cheap failure, and finding linkages across the traditional business clusters. The spin out's CEO operated as a boundary-spanner facilitating the flow of information between different groups and enabling 'communication and understanding to take place across … knowledge domains' (Taylor and Helfat 2009, p. 721). In his own words, 'amazing things happen at intersections.' The consolidation of the spin out into the firm's consulting arm also highlights the importance of driving the innovation culture deeper into the exploit landscape. The case reinforces the theoretical position (Leana and Barry 2000; Luscher and Lewis 2008; Davis et al. 2009) that one way to take advantage of the explore—exploit tension involves the use of heterogeneous, modular groups and units capable of working in the gray area, where the normal rules and expectations become more elastic, and novel propositions can be tested in the real world quickly at a low cost and risk. However, our model suggests that the tension becomes most productive when the groups move back into the fold.

From Explore—Exploit Tension to Organizing Forms Action

Theorists such as Baghai et al. (2000) and Moore (2007) have warned that attempting to optimize exploitation and exploration does not work. Balancing the two or trying to opportunistically shift resources between them paradoxically tends to leave investment in innovation for plentiful times when it is least needed. Along with the resource allocation issue, leaders and managers need to continually switch metrics and priorities around time horizons, performance, and investment outcomes. According to dualities theory, the key revolves around maintaining a deliberate disequilibrium where everyone accepts uncertainty, tension, ambiguity, and the real possibility of deleterious performance ratings in order to ensure that the next stage of the company's life cycle will be successful. In the real, cut and thrust world of daily business, however, the duality mindset remains difficult to enact. The firm's response held consistent with a dualities approach, but as reported as a first theme, the key unit of thinking shifted to a fluid organizing mindset.

Where researchers like Tushman et al. (2012) proposed that organizations create numerous and blurred boundary options, the firm's boundaries solidified through the use of fluid organizing forms. Clusters delivered the daily bread but seized virtually 'free' ready-made new products, mindful that their performances were being measured not just by quarterly profits, but also by a steady turnover of revenue sources. A solid foundation of stability serves as 'both an outcome and medium of change' (Farjoun 2010, p. 203). Our data indicated that management support, work discretion, organizational boundaries, rewards and reinforcements, and time availability were all fluid and conditional upon the leader and his or her project. That is, they were temporarily molded to the indigenous needs of the project at hand. The firm's H2 project initiatives were more successful when they avoided prescriptive, formulaic impositions and simply worked hard to support those that were going well. A benefit of the dualities mindset enacted with these organizing forms was the disinclination to predict the unpredictable and just get on with testing the market.

Fluid Structuring for Ambidexterity Capability

In framing their challenge, the firm's leadership faced the twin and often contraindicated necessities of a controlled, commercial yield alongside radical product innovation. Between what they described as an 'explore—exploit' tension, the firm responded by creating novel structures and initiatives characterized by three organizing forms described in Fig. 5.1 and outlined earlier as themes. The first tier combined a precisely executed cluster business model dedicated to surgical measurement, control, efficiency, and incremental improvement, with an innovation program committed to rapid prototyping and concept proofs in the market instead of lengthy commercialization plans (plus a design-oriented, user-based mode of thinking about client experiences instead of off-the-shelf service solutions). The second tier focused on discrete mid-horizon (H2) commercialization projects seeking either to inject modified products into existing clusters as rapidly as possible, or to work toward the development of radical, 'game-changing innovations.' The third tier involved a pseudo spin-out, spin-in, spin-out model of ongoing start-ups designed to operate freely as entrepreneurial entities that brought their newfound products and culture into the main operations. These three tiers offer some clues toward stimulating an environment conducive to ambidexterity capability.

Driving Innovation and Control with Duality Tension

Instead of relying upon change to the engrained performance mindsets of cluster partner leaders, the firm's executive leadership treated the entire firm as an organizing sandpit. They constructed an innovation program open to all staff, which in turn fed incremental innovations into the clusters while releasing the potential of higher risk, disruptive ideas to connect with a new business unit that could deal with them. At the same time, the business unit clusters became further 'granularized.' Paradoxically, the clusters loosened their stranglehold on the past and opened the door to new possibilities, driven by a newfound cohort of intrapreneurs keen to exercise some entrepreneurial thinking without cash-based cluster investment and without compromising day-to-day measures.

Paradox of Location: Exploring Radical Innovation/Exploiting Existing Initiatives

As a second tier of fluid organizing forms, the firm introduced horizon two projects designed to identify high-potential ideas and rapidly accelerate their commercialization. Supported by a small central unit, H2 project leaders linked in with parent clusters or with the spin-out. The result was a series of new product/service prototypes, tested in the market, and converging upon a deliverable commercial offering.

CONCLUSION

This case revealed that part of the firm's success lay with their fluid organizing forms approach to dealing with the explore—exploit tension. Instead of seeking to delimit it, they sought its escalation into a productive tension, sufficiently powerful to impel individuals to innovate, but sufficiently contained to be captured at an organizational level. According to the literature, a core challenge in reconciling the tension inherent in managing innovation and commercialization structures pivots around a misalignment between the organization and the individual. For example, there has been an overreliance on developing the organizational mechanisms needed to enable dualities with little appreciation or understanding for the importance of individual fluidity (Raisch et al. 2009). Individuals tend to find it difficult to excel at both exploitation and exploration; they must manage contradictions and conflicting goals, work with uncertainty and ambiguity, be comfortable taking risks, and perform diverse, swiftly changing roles. We found, in the firm's case, a series of organizing forms in which individuals can be innovative *and* stable. Such an environment appears to swiftly encourage and propel ambidexterity capabilities. In the following chapter, we take an in-depth look at the spin-out firm introduced here in order to tangibly connect fluidity with ambidexterity capabilities.

REFERENCES

Baghai, M. A., Everingham, B., & White, D. (2000). Growth down under. *The McKinsey Quarterly, 1*(1), 12–14.

Berger, P., & Luckmann, T. (1967). *The social construction of reality: A treatise in the sociology of knowledge.* Harmondsworth: Penguin Books.

Davis, J. P., Eisenhardt, K. M., & Bingham, C. B. (2009). Optimal structure, market dynamism, and the strategy of simple rule. *Administrative Science Quarterly, 54*(3), 413–452.

Daymon, C., & Holloway, I. (2002). *Qualitative research methods in public relations and marketing communications.* London: Routledge.

Deetz, S. (1996). Crossroads—Describing differences in approaches to organization science: Rethinking Burrell and Morgan and their legacy. *Organization Science, 7*(2), 191–207.

Denis, J.-L., Lamothe, L., & Langley, A. (2001). The dynamics of collective leadership and strategic change in pluralistic organizations. *Academy of Management Journal, 44*(4), 809–837.

Eisenhardt, K. M. (1989). Building theories from case study research. *Academy of Management Review, 14*(4), 532–550.

Farjoun, M. (2010). Beyond dualism: Stability and change as a duality. *Academy of Management Review, 35*(2), 202–225.

Galvin, P. (2014). A new vision for the Journal of Management & Organization: The role of context. *Journal of Management & Organization, 20*(1), 1–5.

Glaser, B., & Strauss, A. (1967). *The discovery of grounded theory*. Chicago: Aldine Publishing.

Isabella, L. A. (1990). Evolving interpretations as change unfolds: How managers construe key organizational events. *Academy of Management Journal, 33*(1), 7–41.

Leana, C. R., & Barry, B. (2000). Stability and change as simultaneous experiences in organizational life. *Academy of Management Review, 25*(4), 753–759.

Luscher, L. S., & Lewis, M. E. (2008). Organizational change and managerial sensemaking: Working through paradox. *Academy of Management Journal, 51*(2), 221–240.

Moore, G. A. (2007). To succeed in the long term, focus on the middle term. *Harvard Business Review, 85*(7–8), 2–8.

Raisch, S., Birkinshaw, J., Probst, G., & Tushman, M. L. (2009). Organizational ambidexterity: Balancing exploitation and exploration for sustained performance. *Organization Science, 20*(4), 685–695.

Strauss, A., & Corbin, J. (1990). *Basics of qualitative research: Grounded theory procedures and techniques*. Newbury Park, CA: Sage.

Taylor, A., & Helfat, C. E. (2009). Organizational linkages for surviving technical change: Complementary assets, middle management, and ambidexterity. *Organization Science, 20*(4), 718–739.

Tushman, M., Lakhani, K., & Lifshitz-Assaf, H. (2012). Open innovation and organizational design. *Journal of Organizational Design, 1*(1), 24–27.

Whiteley, A. M. (2004). Grounded research: A modified grounded theory for the business setting. *Qualitative Research Journal, 4*(2), 27–47.

CHAPTER 6

Breaking Out

Abstract This chapter examines a case study focusing on ambidexterity capabilities. The leadership in this chapter's case firm made a bold commitment to positioning innovation as a core feature of its culture. An experimental approach was adopted that heightened the explore—exploit tension, propelling the firm's innovation objectives. They iteratively developed an innovation mindset designed to stimulate new, 'ahead of the curve,' services in order to secure a sustainable turnover of new revenue. The chapter emphasizes the case firm's formation of an agile, technology-focused, product-oriented business unit that turned the traditional services firm business model upside-down by using low touch services, remote connections, self-service clients, and a high volume of transactions and clients, all within more modest margins. The chapter presents the firm's response as a form of ambidexterity capability.

Keywords Spin out · Experimentation · Agile · Technology

INTRODUCTION

In our previous chapters, we emphasized that for firm innovation to become more than a rhetorical by-line, traditional forms of organizing may not be the most effective approach to harnessing the inherent tension between exploit and explore endeavors. The longitudinal case study at the center of this chapter speaks to a more fluid form of organizing that better underpins ambidexterity capabilities. The leadership in this chapter's case firm, when faced with what they described as 'a burning platform,' made a bold commitment to position innovation as a core feature of their culture. At the same time, they had learnt the imperative of maintaining a controlled, commercial yield from the firm's mature and well-respected services. Between this explore—exploit tension, the firm experimented. They iteratively developed an innovation mindset designed to stimulate new, 'ahead of the curve,' services in order to secure a sustaining turnover of new revenue. Core to this experimental foray into the uncertainty of innovation was the creation of 'TechFirm.' TechFirm took the form of an agile, technology-focused, product-oriented business unit that turned the traditional services firm business model upside-down by using low touch services, remote connections, self-service clients, and a high volume of transactions and clients, all within more modest margins.

This chapter charts the life cycle of TechFirm. While there were swings and roundabouts in the way TechFirm morphed and developed, at the heart of this rather incongruous approach was what the founding CEO described as an ideology that 'innovation doesn't have to be complicated, you just have to find ways to go around the road blocks.' From a practical perspective, this case demonstrates the importance of heterogeneous capabilities leveraged by fluid groups that engage a co-design approach to product/service development centering on rapid, 'low-fi' prototyping undertaken in collaboration with lighthouse clients. TechFirm illustrates how a firm can span the troublesome implementation gap between theoretical notions of explore—exploit and the successful delivery of both innovation and commercial efficiency. Put bluntly, TechFirm *is* ambidexterity capability.

Making Innovation Pay

The CEO of our case firm held no illusions regarding the structural and managerial tensions that accompanied a commitment to innovation through both exploitation and exploration. He also knew very well that standing on a 'burning platform' would not only drive the company toward further disaster, it would most likely result in a significant and irreversible erosion of the firm's existing client base. The problem at hand was one of *how* to deal with such contradictory forces. After all, according to previous studies, high levels of exploration and exploitation at the same time means dealing with completely different structures in a business as 'mutually enabling constituent' parts (Farjoun 2010, p. 205). Moreover, most firms '…are not generally equipped to cope with fragmentation and high ambiguity' (Seo et al. 2004, p. 162). For the most part, successful businesses in the professional services sector reside in the exploitation domain and rely upon long-standing, conservative, and predictable products and services based on trust, efficiency, and reliability.

In this chapter, we examine previously highlighted exploit—explore challenges through the TechFirm lens. We detail how the business unit evolved, was spun out into a successful independent operation, and then ultimately was absorbed back into a core service line, effectively ending its independent life. At the same time, we chart TechFirm's specific and real-world experience against the more generic, theoretical advice found in previous studies and commentaries. This case study, like the ones we explored in the two preceding chapters, was developed from a wide-ranging series of in-depth interviews with TechFirm's members, the senior executive of its parent firm, and a purposive sample of partners from across the eight service arms who interacted with TechFirm. Supporting these interviews, ethnographic data were collected via immersion in TechFirm through a series of participant observations undertaken over more than 36 months.

As a result of meticulous and extensive data collection and analysis, pivotal relationships and themes were identified. Findings revealed that the TechFirm initiative fast-tracked structural adaptation, fostering ambidexterity capability across what had always been conservative service units. In practice, TechFirm's model as a spun-out/spun-in unit worked in an osmotic fashion, innovating at the core while inspiring at the edges (Gilbert et al. 2015). From a theoretical perspective, we conclude that the use of TechFirm as an innovation catalyst resulted in a practical and effective realization of ambidexterity capability resulting from a firm deliberately harnessing explore—exploit tension. By avoiding an opportunistic, dynamic oscillation between innovation and efficiency, this case illustrates how the two can be pursued together through structured innovation.

STRUCTURING INNOVATION: EVOLUTION AND REVOLUTION

TechFirm's story began as early as 2004 when an internal innovation initiative was established at the case firm, designed to provide employees with a forum to share and generate ideas. The program's governance model comprised an appointed executive under which an Innovation Council was established to assess ideas, solicited from all members of the firm and of which there was no shortage. Many of the ideas put to the Innovation Council suggested replacing traditional face-to-face delivery of client services with digital service delivery. The proposals highlighted how traditional services could be delivered online more effectively and inexpensively while at a higher quality. Another common theme involved adding an online component to a traditional client service that would offer a better end-to-end experience. The concept for TechFirm was spawned in the confluence of these early forays into online service delivery.

Despite the enthusiasm for new ways of doing things, it took until 2008 for TechFirm to become formally established with the mandate to deliver professional services digitally in order to provide clients with easy access to a range of financial tools, training modules, compliance, and human resource solutions. Specific examples included accounting, benchmarking, education around innovation and leadership, identity verification for the banking sector (e.g., for 100-point checks), and social media services. The new online products illustrate the shift from a traditional compliance mindset to one of business adviser. Under the new model, both accountant and client could access real-time data from different geographic locations, all automatically synchronized and updated. It was, according to a senior TechFirm executive, 'a world of joy' for the accountants.

TechFirm evolved rapidly using a technology-driven, service/product hybrid underpinned by a competitive model emphasizing rapid prototyping, proof of concept, fast and cheap failure, and an agile innovation pipeline system. As a result, TechFirm became a catalyst for shuttling existing, but siloed capabilities into the innovation 'engine.' Here designers, coders, application specialists, programmers, and data visualization experts were able to transform conventional services into new products. Furthermore, TechFirm partnered with 'lighthouse' clients to co-design and test their innovations in the foundry of real-world demands, constantly experimenting and upgrading based on user feedback.

The start-up approach taken by TechFirm prospered using a unique combination of two offerings. First, a consulting model focusing on the Internet, intranet and multichannel strategy, customer experience design, social media strategy, mobile applications development, and digital technology architecture and implementation. Second, it wielded stand-alone technology products to supplement existing services or stimulate new consulting opportunities. A central ingredient in TechFirm's success was a 'can do' mindset, bolstered by a bespoke innovation pipeline program and a leader for whom innovation equated to energy. As the Chief Strategy Officer noted, TechFirm's CEO '...has always been a voice in the space, still a voice in the space, has always been the advocate that was pushing the boundaries just on the side and that contributed a lot to the positioning that we just had in general in the marketplace, just the profile that a guy like he had.' Indeed, for TechFirm's CEO, the vision was to be 'breathtaking in execution.' With digitization a core part of the future for professional services firms, TechFirm aimed to be a 'game changer' and global pioneer in online professional services delivery.

CAPTURING INNOVATION

Following its inception in 2008, TechFirm adopted what was then a new-to-market enterprise social networking platform, widely known today as Yammer, to solicit new ideas from anywhere in the firm. Within Yammer, the ideas coalesced under general themes such as social media, mobile applications or gamification. Combined with a small, group-based Innovation Café process for collective brainstorming incorporating contributors from diverse backgrounds, expertise, and positions in the firm, the pool of new ideas was next subjected to an online game mechanic for rating. Using such an online innovation idea capture tool enabled votes to be rationed, thus reducing the sweeping range of ideas to a more targeted and manageable handful. The Council next looked for game-changing ideas. Among the most compelling were those ripe for immediate commercialization in order to generate revenue over the forthcoming 12 months, or improve the company's agility through streamlined processes. TechFirm's CEO observed, '...the effect of the process is to create, generate ideas, submit those ideas in a transparent way, open through our innovation idea capture tool, and then people promote it through our internal social media networks and through their mates and all that stuff and try and get them to vote for it, comment on it, collaborate around it.' The innovation idea capture platform operated as a permanent program regularly supplemented with themes, festivals, and cafés that kept the innovation fires burning.

Ideas were reviewed every two weeks by the Council, which subsequently allocated micro-funding to the most promising based on votes and perceived alignment with existing offerings. The micro-funding model utilized a venture capital philosophy but with an unusual twist. Once an idea had been selected, its originator was allocated AUS$10,000 on the basis of a one-page pitch: 'Put the idea in, we'll review it within two weeks, if we like it we'll give you a micro-fund of $10,000, which gives enough oxygen to the idea to get it going. We help navigate within the organization as to who the best people are to help you get that idea up and running, prototype it, test it, if there's an appetite for it, expand it.' The micro-fund however, in a departure from the venture fund approach, worked on the basis of a 'pool of time' concept instead of actual dollars. Under the concept, a dollar value was allocated against every project. As a result, project champions could work on their ideas up to the value of AUS$10,000 without compromising their service line revenues, because the project was treated in the same manner as a client billing process. As a method of cyclical regulation to modulate the flow, the Council could modify the realization percentage on the pool of time. If new ideas were overflowing, for example, realization on the time could be decreased so that project champions work for less than 100% of full billing rates. Innovators could then receive up to AUS$50,000 in additional bootstrapping finance provided their six-weekly reports demonstrated proof of concept, client interest, and ultimately a commercial case. Projects were broken into discrete chunks or 'staged gates,' where exits could be taken discontinuing, divesting, spinning in, or commercializing the new product or service.

Although simple, the genius behind the pool of time model was that it protected partners' performance measures—linked of course to exploit metrics—thereby encouraging quick forays into innovation, as its costs did not come in either implementation time or preparing lengthy business cases. Underpinning the model, TechFirm's CEO insisted on prioritizing action over planning: 'So it's much more about prototyping, bringing the idea to life, socializing it, finding somebody who wants to run with it as opposed to going and plan, plan, plan, plan. And that's why I don't believe in doing business cases around innovation. I believe in doing prototypes.'

Studies by Brown and Eisenhardt (1997) some 20 years ago add weight to this action over planning approach. They observed that successful firms balance structure and chaos, and rely on a range of low-cost experimental initiatives as forays into the future. Paradox-consistent thinking proved more effective than planning for, or reacting to, unforeseen changes. Smith and Tushman (2005) similarly concluded that so-called paradoxical thinking was a key enabler in the simultaneous pursuit of exploration and exploitation. The TechFirm CEO perhaps would not agree that innovation requires paradoxical thinking, however. To him, innovation is neither paradoxical nor complicated.

Exploiting Exploration

One pivotal theme emerging from the case data—also relevant to the case in Chap. 5—reflects the significant challenge in reconciling a misalignment between the organization and its individual employees. Research has focused intensely on macro-organizational ambidexterity capabilities, but has largely neglected the individual's proficiencies. It remains a major challenge for individuals to be effective at both exploiting and exploring, particularly when there is a lack of clarity around expectations, roles, and rewards. Some studies have tried to identify the elements of individual ambidexterity. For example, Raisch et al. (2009) concluded that the key attributes include: (1) being party to effective top-down/bottom-up and lateral knowledge flows; (2) the ability to consider both short- and long-term issues; and (3) a robust prior-related knowledge and experience base that fuels curiosity and facilitates new knowledge. While this suite of features appears simple, locating the right candidates or training new leaders to operate with 'ambidexterity' is easier in theory than practice. However, TechFirm demonstrated a new organizing mode wherein individuals exercised unusual creativity. In many ways, TechFirm exemplified theoretical ambidexterity in action, as it operated in the nexus between the control paradigm of a conservative exploit-driven firm and the open, unrestrained mindset of the entrepreneurial, explore-driven start-up business unit.

Previous research tends to show that even with the right people, organizational structures, systems, and processes all constrain ambidexterity thought and action (Ghoshal and Bartlett 1995; Raisch et al. 2009). After all, most large firms need control and efficiency to continually deliver on their profit targets. As a result, they must find the right balance of control mechanisms: executive planning, performance measurement, reward systems, and client delivery processes, while simultaneously encouraging the freedom to pursue new innovations (Jansen et al. 2006; Lubatkin et al. 2006). An example can be seen through Andriopoulos and Lewis' (2009, p. 707) comparative case study of five ambidextrous firms, which underlined the importance of both integration and differentiation as 'powerful, complementary tactics for fostering ambidexterity.' Similarly, Eisenhardt and Brown (1998, p. 703) found that organization leaders needed to leverage paradox 'in a creative way that captures both extremes.' But, what does it mean to 'leverage paradox' in a real firm facing a poor pipeline of innovative new products and services where the only consistent revenue depends on exploiting existing services? For our case, the answer was to drive toward greater acceptance and practice of innovation. At the same time, TechFirm modeled new ways of interpreting, acting, and recognizing success.

MANAGING FOR EXPLORE AND EXPLOIT: PRACTICE AND THEORY

The explore and exploit activities of TechFirm were necessarily managed quite differently from its parent firm, although the tensions between the two forms of activities were also instrumental to the unit's success. The exploit components were managed according to traditional balance sheet criteria, with consideration given to gross margins, profit and loss, salary costs, and other control and efficiency-related metrics. While TechFirm's CEO was responsible for delivering on all performance metrics, he focused more on nurturing explorative ventures in their infancy, which held potential for future profit. However, TechFirm's CEO, like any other, could not escape the significant pressure to demonstrate value; an imperative exacerbated by the absence of tangible metrics or balance sheet credits around exploration that other partners either recognized or used to measure performance in their own exploit-based business units. Nascent 'explore' initiatives required different benchmarks and performance measures from their established 'exploit' counterparts. Unless such alternative requirements received recognition, pioneering initiatives were likely to be subsumed or extinguished because they failed to demonstrate 'value' via conventional measures of success.

TechFirm's CEO addressed the measurement dilemma by taking a 'fame, fortune, and franchise' approach. His aim was to make innovation an undeniable force and was therefore careful not to undertake any 'pioneering initiatives' in the absence of an end-user: '…I'll never prototype stuff in an echo chamber … the first thing we do with a prototype is … let's get four or five potential customers in there and say this is what we're thinking … Does anybody want to work with us on this?' Testing new ideas with current or prospective clients became an essential step: 'I think doing any innovation in the absence of somebody who's going to actually use or benefit from the service product or experience that you deliver is crazy. So I want a real life end-user.' He added, 'share early, share often before it's ready and see if somebody wants to go on the journey with us, you know, customer reference groups we use a lot, lighthouse client is another term we use.' TechFirm's model addressed the explore—exploit tension by maximizing speed to market while mitigating cost and risk, all the while shaping a compelling narrative around innovation.

Leveraging TechFirm's explore—exploit tension brought into play some of the constraining issues that innovation theorists such as Baghai et al. (2000) and Moore (2007) have highlighted. Such constraints appear when organizations attempt to calibrate strategy and operational systems around an (imperfect) 'optimization' of exploitation-exploration. Simply balancing the two out in equal but modest amounts does not work. Caution about the challenges that accompany resource cannibalization seems warranted. After all, it is not so easy for companies to pour considerable resources into both explore and exploit at the same time. Indeed, as Gilbert et al. (2015) observed, when such concurrent plays do not succeed, or even face minor set backs, firms revert to what they know in favor of what they might come to know. Moreover, resource migration rarely occurs from exploit to explore, particularly given the variance in cycle times. Managing in the exploit domain dictates that systems and processes are geared around budgeting and reporting over a fiscal year (particularly so if shareholder wealth-building is the primary focus of an organization). The convention is steadfastly predicated upon the almost unassailable notion that core business can be predicted, and thus efficiency gains can be achieved through incremental system and process improvements.

Few managers are comfortable with uncertainty and ambiguity, especially when it threatens profit margins and performance bonuses. Little wonder many avoid the risks that go with innovation. Equally, commercializing creativity in the explore domain demands alternative metrics and priorities around time horizons, performance, and investment outcomes. Sustaining the balancing act requires management and organizational buy-in of the highest order, as well as no small measure of comfort with ambiguity. Part of the trick TechFirm mastered in dealing with the explore—exploit tension revolved around maintaining a deliberate disequilibrium. To achieve higher levels of ambidexterity capability, TechFirm increased service value and revenue growth through incremental innovations across core activities in the firm, consequently unlocking creativity, competency, and diverse capacities at the edges. They also worked to infuse innovation as an essential part of the firm's broader cultural renewal. In turn, this cultural shift seeded more radical experimentation along with a stronger appetite for risk.

In the theoretical world, TechFirm's approach mirrors calls to abandon debates about open versus closed organizational boundaries. Instead, design researchers like Tushman et al. (2012) have proposed a more complex interpretation where organizational boundaries are blurred because firms simultaneously pursue numerous boundary options, including some involving closed vertical integration, some in the middle ground like strategic alliances, and some open innovation: 'The simultaneous pursuit of multiple types of organizational boundaries results in organizations that can attend to complex, often internally inconsistent, innovation logics and their structural and process requirements' (Tushman et al. pp. 24–25).

Similar thinking can be found in other studies on ambidextrous organizations. For example, Andriopoulos and Lewis's (2009) research emphasized integrative efforts. In fact, the presence of *integration*, promoting the coordination of exploit—explore efforts through what they called synergistic interdependencies between apparent opposites, proved advantageous to performance. It worked through increased 'absorptive capacity' among organizational members. A little of each will not do. Similarly, TechFirm performed better under conditions where both explore and exploit received focused attention. For example, while exploitative activities aim to convert knowledge into marketable products and services, an organization's bank of knowledge, skills, and expertise weaken and decline without the exploration of new ideas. Correspondingly, exploration helps build new knowledge and skills, but its potential may not be fully realized without exploitation efforts.

Another pertinent study by Taylor and Helfat (2009) pointed to the importance of 'linkages'—creating connections through communication and coordination mechanisms that stimulate transparency, trust, and open two-way information channels between explore and exploit areas of an organization. TechFirm's CEO and his team became key connectors, acting as boundary spanners facilitating information flows between different groups and enabling 'communication and understanding to take place across ... knowledge domains' (Taylor and Helfat, p. 721). Collaborative and synergistic activities played a central role in success. In the words of TechFirm's CEO, 'Amazing things happen at intersections.' As Andriopoulos and Lewis (2009, p. 708) put it, exploration and exploitation represent two modes of innovation that are 'mutually reinforcing.' Leaders must be able to think and act 'in broad and integrative ways' (Chen and Miller 2010, p. 9).

As this case has exemplified, innovation can occur within a context of rigid structures ensuring accountability and due process, but also where performance relies upon agility. Other innovation and change cases similarly reveal that growing firms rely on an interactive mix of continuity (exploitation) and change (exploration) (Leana and Barry 2000; Luscher and Lewis 2008; Davis et al. 2009). As boundaries blur and organizations operate deeper in global markets, 'loose-tight' relationships become more important, where structures enhance responsiveness while delivering efficiency. In short, as we have described in this case, freedom needs boundaries while boundaries need spanning.

Opportunities for innovation cannot transpire at the expense of control. A solid foundation of stability serves as 'both an outcome and medium of change' (Farjoun 2010, p. 203). It provides the solid base from which innovative ventures, critical for both renewal and longevity, can proceed. The challenge, as the TechFirm case has highlighted, lies in determining how to excel at both, simultaneously maximizing performance efficiencies while creating an adaptive, responsive, innovation-driven culture. Moreover, in a world where successful innovation means building on a treadmill, nothing can be more certain than imminent change and the need to run faster to stay in the same place.

In 2012, TechFirm reintegrated with its parent company's Management Consulting Practice. The move aimed to accumulate a critical mass of services and capabilities behind the innovation agenda. For example, both product and market opportunities expanded radically when TechFirm's innovation capability was leveraged with consulting expertise to share services and build offerings in partnership with other parts of the firm. TechFirm had been spun out in order to provide the necessary freedom or *right* to innovate. By securing its mandate and then being spun back in, TechFirm had shifted the entire firm's focus to having the *responsibility* to innovate. Innovation needed freedom until it became sustainable, then needed control to generate scale in order to deliver a return on investment.

STRUCTURING INNOVATIVE TENSION: LEARNING FROM THE TECHFIRM LIFE CYCLE

The notion that competing forces permeate the practical and philosophical world enjoys a long tradition, from Hegel's (1899) original dialectic to Smith and Lewis's (2011) 'dynamic equilibrium.' Acknowledging the presence of tension has not been the problem. Rather, most organizations face a daunting 'implementation gap,' which to date has not been filled with prescriptive advice from theorists. Smith and Lewis's (2011) dynamic equilibrium model of organizing represented a recent attempt to find some answers to the implementation problem. As they explained, a dynamic equilibrium assumes constant motion across opposing forces. How practical their dynamic equilibrium model will prove for practitioners has yet to be determined, as it again relies significantly on a highly theoretical proposition. The lack of practical implementation frameworks reflect the difficulty in the day-to-day management of explore—exploit tensions. Evans et al. (2011) for example drew on the navigator/helmsman metaphor to underscore the complexities and highly skilled nature of the task. Like the helmsman on a ship, organizational managers must negotiate the unrelenting, variable tension between the need to steer a particular course while playing heed to shifting winds and currents (Evans et al. 2011, p. 74). Beyond metaphors and conceptual constructs, what options do managers have for conceptualizing explore—exploit tensions while practically triggering innovation at a time when efficiency has never been more important?

As discussed extensively in Chap. 2, we stressed the importance of developing ambidexterity capability, however messy it might seem. Jackson and Harris's (2003) depiction of ambidextrous forms as a 'mix and match' approach reflects the (albeit untidy) utility of ambidextrous tension. Other researchers have found support for a mix of organic, mechanistic, and knowledge-based structures (e.g. Wang and Ahmed 2003). The TechFirm case demonstrated such a mix, although its deployment was more precise than an experimental 'mix and match' approach implies. Chapter 2 referred to the more structured version of ambidextrous design advocated by Raisch et al. (2009) who, as we recorded earlier, identified four organizing tensions: differentiation/integration; individual/organizational; static/dynamic; and internal/external. All of these can be seen to some extent in this case study, including the firm's careful experimentation with modular design structures, where a module operated as '...a unit within which there is robust interaction among structural elements but superfluous interaction with elements in other units' (Baldwin and Clark 2000, p. 63). Again, however, recognizing tensions does not translate seamlessly into practical innovation. Nor was it just a matter of adding new modules of activity to work contiguously with conventional business units. In theory, modular forms of organizing represent an offspring from ambidextrous forms, characterized by resilience, flexibility, and responsiveness through the ability to restructure swiftly in dynamic and uncertain environments (Schilling and Steensma 2001; Galunic and Eisenhardt 2001). Like the notion of 'architectural ambidexterity,' modularity concerns, '...structures based on minimizing interdependence between modules and maximizing interdependence within modules' (Ethiraj and Levinthal 2004, p. 432). Going a step beyond the recommendation for modular business units that rely on internal fluidity, research conducted by Tushman et al. (2010) found correlations between performance and heterogeneity within structures. Just as we observed in TechFirm, they advocated for 'highly differentiated and inconsistent' structures (Tushman et al. p. 1336).

CONCLUSION

Ambidextrous and modular approaches recognize that attempts to 'optimize' both exploitation and exploration at the same time do not work. Instead, theorists such as Tushman et al. (2010) and Farjoun (2010) have proposed that organizations must oscillate dynamically between the two. The choice to press ahead aggressively in one or the other direction can come in response to environmental volatility and other pressures or opportunities. Little advice is offered, however, on how to perform 'dynamic oscillation' in practice, especially since most managers would find it challenging to nominate the circumstances under which either efficiency or innovation should assume a lower priority. As a consequence, we diverge from the oscillation strategy in favor of building dualities through ambidexterity capabilities: not one or the other, but both at the same time.

Some lessons for building ambidexterity capability—and by consequence—designing innovation may be seen through the TechFirm case. A first necessary condition involves strategic intention. Leadership needs to make a commitment around agile, modular, and even experimental structures and groupings. As the TechFirm CEO suggested, 'things happen' at the intersections. Sculptured slogans may claim a commitment to innovation, but if unaccompanied by the deeds necessary to embed innovation in the fabric of an organization, management will fall back upon the predictable, safe, and modest returns of incremental efficiency. Further, a tangible set of new practices and structures must underpin the claim for innovation. While novel, modular business units need to be free to operate responsively in partnership with 'lighthouse' clients and with an emphasis on rapid, 'fast fail, fast scale' methods, leaders must be willing and able to assume the risk of allowing and indeed encouraging freedom. It also requires a significant investment in mobilizing support around what can be viewed as ideas from the 'edges.' Even more challenging, leaders must convince pivotal business unit gatekeepers that *more* uncertainty and ambiguity is desirable.

We have reported how the TechFirm case exemplified the controlled importation of looser, non-traditional structures. Instead of an innovation team or department, diverse and previously partitioned capabilities were brought together in agile enterprise teams. Combined with the pool of time method, innovation was seeded from the core to the edges without threatening the existing profit lines of business units where immediate sales reign sovereign over stoking the pipeline. The spin-out–spin-in model was so successful that the approach is being repeated at the case firm with another innovation-centered business unit. It has been given unprecedented freedom to explore the intersections between emerging technologies and their implications for new models of business, for example, in realizing the potential of a future cashless society. In the following chapter, we take a final step in making tangible the dual organization and what can be done in practice to build ambidexterity capabilities.

REFERENCES

Andriopoulos, C., & Lewis, M. W. (2009). Exploitation-exploration tensions and organizational ambidexterity: Managing paradoxes of innovation. *Organization Science, 20*(4), 696–717.

Baghai, M. A., Everingham, B., & White, D. (2000). Growth down under. *The McKinsey Quarterly, 1*(1), 12–14.

Baldwin, C., & Clark, K. (2000). *Design rules: The power of modularity*. Cambridge, MA: MIT Press.

Brown, S. L., & Eisenhardt, K. M. (1997). The art of continuous change: Linking complexity theory and time paced evolution in relentlessly shifting organizations. *Administrative Science Quarterly, 42*(1), 1–34.

Chen, M.-J., & Miller, D. (2010). West meets East: Toward an ambicultural approach to management. *Academy of Management Perspectives, 24*(4), 17–22.

Davis, J. P., Eisenhardt, K. M., & Bingham, C. B. (2009). Optimal structure, market dynamism, and the strategy of simple rule. *Administrative Science Quarterly, 54*(3), 413–452.

Eisenhardt, K. M., & Brown, S. L. (1998). Competing on the edge: Strategy as structured chaos. *Long Range Planning, 31*(5), 786–789.

Ethiraj, S., & Levinthal, D. (2004). Bounded rationality and the search for organizational architecture: An evolutionary perspective on the design of organizations and their evolvability. *Administrative Science Quarterly, 49*(3), 404–437.

Evans, P., Pucik, V., & Bjorkman, I. (2011). *The global challenge: International human resource management* (2nd ed.). New York: McGraw-Hill Irwin.

Farjoun, M. (2010). Beyond dualism: Stability and change as a duality. *Academy of Management Review, 35*(2), 202–225.

Galunic, C. D., & Eisenhardt, K. M. (2001). Architectural innovation and modular corporate form. *Academy of Management Journal, 44*(6), 1229–1249.

Ghoshal, S., & Bartlett, C. A. (1995). Building the entrepreneurial corporation: New organizational processes, new managerial tasks. *European Management Journal, 13*(2), 139–155.

Gilbert, D. H., Smith, A. C. T., & Sutherland, F. (2015). Osmotic strategy: Innovating at the core to inspire at the edges. *Organizational Dynamics, 44*(3), 217–225.

Hegel, G. W. F. (1899). Trans. by J. Sibree. *Philosophy of history.* New York: The Colonial Press.

Jackson, P., & Harris, L. (2003). E-business and organisational change: Reconciling traditional values with business transformation. *Journal of Organisational Change Management, 16*(5), 497–511.

Jansen, J. J. P., van den Bosch, F. A. J., & Volberda, H. W. (2006). Exploratory innovation, exploitative innovation, and performance: Effects of organizational antecedents and environmental moderators. *Management Science, 52*(11), 1661–1674.

Leana, C. R., & Barry, B. (2000). Stability and change as simultaneous experiences in organizational life. *Academy of Management Review, 25*(4), 753–759.

Lubatkin, M. H., Simsek, Z., Ling, Y., & Veiga, J. F. (2006). Ambidexterity and performance in small-to-medium-sized firms. *Journal of Management, 32*(5), 646–672.

Luscher, L. S., & Lewis, M. E. (2008). Organizational change and managerial sensemaking: Working through paradox. *Academy of Management Journal, 51*(2), 221–240.

Moore, G. A. (2007). To succeed in the long term, focus on the middle term. *Harvard Business Review, 85*(7–8), 2–8.

Raisch, S., Birkinshaw, J., Probst, G., & Tushman, M. L. (2009). Organizational ambidexterity: Balancing exploitation and exploration for sustained performance. *Organization Science, 20*(4), 685–695.

Schilling, M. A., & Steensma, H. (2001). The use of modular organizational forms: An industry-level analysis. *The Academy of Management Journal, 44*(6), 1149–1168.

Seo, M. G., Putnam, L. L., & Bartunek, J. M. (2004). Dualities and tensions of planned organizational change. In M. S. Poole & A. H. Van de Ven (Eds.), *Handbook of organizational change and innovation* (pp. 73–107). Oxford, UK: Oxford University Press.

Smith, W., & Lewis, M. (2011). Toward a theory of paradox: A dynamic equilibrium model of organizing. *Academy of Management Review, 36*(2), 381–403.

Smith, W. K., & Tushman, M. L. (2005). Managing strategic contradictions: A top management model for managing innovation streams. *Organization Science, 16*(5), 522–536.

Taylor, A., & Helfat, C. E. (2009). Organizational linkages for surviving technical change: Complementary assets, middle management, and ambidexterity. *Organization Science, 20*(4), 718–739.

Tushman, M., Lakhani, K., & Lifshitz-Assaf, H. (2012). Open innovation and organizational design. *Journal of Organizational Design, 1*(1), 24–27.

Tushman, M., Smith, W. K., Chapman Wood, R., Westerman, G., & O'Reilly, C. A. (2010). Organizational designs and innovation streams. *Industrial and Corporate Change, 19*(5), 1331–1366.

Wang, C., & Ahmed, P. (2003). Structure and structural dimensions for knowledge-based organizations. *Measuring Business Excellence, 7*(1), 51–62.

CHAPTER 7

Designing Business Innovation

Abstract This chapter examines a case study focusing on the practical transition of innovation 'wow' to practical 'how.' It establishes some concrete recommendations for developing ambidexterity capability based on a combination of case lessons, results from salient research, and experience in melding the two. The chapter connects design and innovation, identifies some of the most useful tools and techniques, and prescribes a series of actions stimulating ambidexterity capability. The case data reinforces the nonlinear nature of innovation. In addition, the chapter witnesses the failure of reductionistic attempts to commoditize design thinking, which tend to produce mechanistic outcomes. However, it also demonstrates how powerful design thinking and its associated methodologies and tools can be in democratizing innovation in a conservative, declining firm.

Keywords Design · Design thinking · Democratizing innovation

INTRODUCTION

The previous chapter presented the TechFirm case and how the deliberate design of looser, non-traditional structures was favored over installing an innovation team or department. We described how heterogeneous and partitioned capabilities were brought together in agile enterprise teams. Over more than a decade, the company housing TechFirm had been positioning innovation as its key differentiator, and as part of its approach to

© The Author(s) 2017
A.C.T. Smith et al., *Reinventing Innovation*,
DOI 10.1007/978-3-319-57213-0_7

creating ambidexterity capabilities. In short, they had been trying to build a dual organization to enable market-leading practices in both explore and exploit domains. This journey led the firm to become both an early adopter and a market leader in the use and dissemination of 'Design Thinking.' Their earlier experiences with innovation reflected that of many others, characterized by what Mark Payne, CEO of Fahrenheit 212, described as getting caught up in the 'wow,' but dismally failing on the 'how.' The alternative path that TechFirm followed combined both the 'wow' and the 'how.' A critical part of their success proved to be a commitment to experimentation with design thinking and the suite of tools that could be transposed into the business environment. While design thinking, defined by Lockwood (2010, p. xi) as 'essentially a human-centered innovation process that emphasizes observation, collaboration, fast learning, visualization of ideas, rapid concept prototyping and concurrent business analysis,' has reached some degree of prominence in product design and development, only more recently has it enjoyed a similar degree of popularity in service design. As we describe, the TechFirm model placed humans at the center of innovation, in the process creating both an object (a product) and an experience (a service) within a professional services environment.

In this chapter, we examine the TechFirm transition from 'wow' to 'how.' At the same time, we establish some concrete recommendations for developing ambidexterity capability based on a combination of our case lessons, results from salient research, and our experience in melding the two. The chapter connects design and innovation, identifies some of the most useful tools and techniques, and prescribes a series of actions stimulating ambidexterity capability.

Design and Innovation

Over the past decade or so, design thinking has received attention both within and beyond the contemporary design field. In particular, management and business education have feted design thinking as a revolutionary approach to changing the way we perceive, articulate, and solve the kind of complex problems that cannot be solved through linear pragmatism or planning. Yet, many business commentators, researchers, and practitioners have ignored design's potential to inspire innovation (Hobday et al. 2011). Moreover, in examples where design has been considered, the predominant focus has been on manufacturing, despite the fact that services dominate in developed and advanced economies. Some evidence has indicated that the way design has been conceptualized in business innovation has been narrow, limited to creative and technical domains (Buchanan 1992). It has also been observed that design thinking represents a failed experiment in business because it has been reduced to a 'linear, gated, by-the-book methodology'; what Nussbaum (2011) referred to as 'N + 1 innovation.'

Our experience from the numerous cases examined in this book—in particular the impact of a spun-out, and then reintegrated, digital innovation business unit—has yielded some practical insights about the circumstances under which ambidexterity capabilities can accumulate. One fundamental revelation was that a reductionist view of design and the associated limitations regarding its potential to facilitate innovation were a function of managers confusing the 'thinking' with the 'tools,' seeing the two as one and the same. Furthermore, we have continually observed that managers plan, lead, organize, and control in ways that exclude innovation. Paradoxically, even when managers aggressively pursue innovation, they tend to push it away with excessive programming and planning. It is a bit like a push for workplace democracy forcing 'empowerment' upon employees. Similarly, innovation cannot be mandated through a strategic document demanding its presence, nor can design urge creativity simply because someone hosts a workshop where participants get to write on post-it notes. Seeking immediate solutions via a design 'toolbox' approach without knowledge and understanding of the thinking behind the tools inevitably produces poor outcomes. Nevertheless, as the worldwide push by governments, businesses, and educational institutions to produce individuals comfortable with complexity and ambiguity exemplifies, there has never been a greater need for design thinking, as well as the tools to transform design thinking into action.

In the previous chapter, we revealed that TechFirm sought to avoid the failures of others—not to mention the calamitous errors of their parent firm—by exposing their design thinking experimentation to external benchmarks and critiques. They began by interacting with thought leaders working at the nexus of design and business. Prominent examples included Verganti (2009) and his work on 'design-driven innovation,' Beckman and Barry (2007) and their framework of design thinking as a learning process, Martin (2009) with his 'knowledge funnel,' and Brown's (2009) 'change by design' approach.

The aim according to TechFirm's CEO was 'not to create another hard to remember and actually useless process.' Instead, he sought a growth strategy based on innovation by using design thinking's potential to build and develop ambidexterity capability across the organization. The strategy positioned TechFirm ahead of the curve, especially when combined with the liberal use of enterprise social networking platforms such as Yammer and game mechanics to bolster social inclusion around the firm's innovation drive. The latter also engaged latent innovation capability by providing a swift vehicle to harness creativity irrespective of where it arose across the entire firm. Design 'thinking' and design 'tools' were combined, shifting the game from providing services to designing experiences—a significant new trajectory. TechFirm's ability to innovate at the core while inspiring at the edges also confirmed that a toolbox approach to complex problem solving and opportunity realization lacks substance without the corresponding thinking behind the tools.

THE 'THINKING' AND THE 'TOOLS': DESIGNING FOR INNOVATION

Perhaps the design thinking triumvirate of *desirability, feasibility*, and *viability* best encapsulated TechFirm's approach to innovation. Whether designing strategy, processes and systems, or products and services, constant reference was given to 'who wants it and what value does it represent to end-users?' (desirability); 'how will you build or shape, then test and deliver what you conceive?' (feasibility); and 'how will your solution be resourced, sustained, and profited from?' (viability). For each element in the treble heuristic, TechFirm started from the concrete and analytical by looking at what did and did not work, then shifted into the abstract, to reframe the problem. Next, they would co-design the end product or experience with end-users to synthesize the possibilities, which were subsequently defined and tested. A convergence would ultimately cluster around what worked and what was practical.

Five imperatives were distinguished in the TechFirm approach to innovation thinking and practice, which they expressed colloquially as: (1) make us care; (2) show us something new; (3) tell us what's missing; (4) highlight what can be changed; and (5) make it tangible. Within the five imperatives, the thinking component clarified what was important to users so that convergence could be better achieved by using the appropriate tools, whether co-designing, rapid prototyping, visualizing, personas, mock-ups, service blueprinting and specification, or business model canvases. In the process, the approach verified to others in the firm that design thinking was fundamentally about human needs. The end-user, whether in the form of customers, staff, shareholders, or key stakeholders, such as suppliers and distributors, were brought firmly to the forefront of the firm's consciousness.

TechFirm had developed an innovation engine uniquely combining two offerings. First, it delivered a consulting model focused on designing excellent customer experiences that employed digital disruption as a differentiator. Second, it offered stand-alone technology products that both supplemented existing services and stimulated new consulting opportunities. Design thinking was embedded in the approach too. It served as the critical link between a bespoke innovation pipeline program and the firm-wide renewal strategy. As a result, the entire firm transformed its mindset from providing professional services to designing and selling experiences. Yet the design component maintained a Spartan simplicity, unencumbered by lengthy planning processes and technocratic documentation. Instead, it emphasized an elegant coordination between design thinking and design tools, like a sports car that drives the way it looks. According to TechFirm's CEO, his approach was 'probably a lot more low resolution than most.' In his view, the most important aspect was not the generation of new ideas, but their conversion into rapid execution. Design thinking had unleashed a latent innovation capacity; the explosion of ideas and activity made an undeniable case for the impact of ambidextrous capabilities. Of course, that did not mean that it was smooth sailing all the way along the journey.

TechFirm's CEO had always been acutely aware of the friction between traditional delivery methods and those he advocated based on the design insights secured from actual users. He reinforced Tim Brown's approach to design thinking, emphasizing *insight, observation, and empathy* in balancing the core requirements of 'valuable innovation': *feasibility, viability, and desirability.* Brown (2009) cautioned against a linear, step-by-step approach to design thinking, instead favoring an interaction of three spaces: *inspiration, ideation, and implementation.* Brown maintained that the tools developed and used by designers could be effectively utilized in business. He added, in addition, tools such as user-centered understanding via ethnographic investigation; brainstorming; mind-mapping/visual-thinking; storyboards, improvisations, and scenarios; and rapid prototyping. Martin (2009) proposed similar tools to assist in finding what he saw as inspiration at the extremes. For example, Martin referred to the edges of unfamiliarity where inter-disciplinary creativity and experimentation productively collide with a diverse set of actors in the innovation ecosystem. Intuition and deduction work in harmony without prejudicing one over the other. Although perhaps overstated, TechFirm's revolution was analogous to becoming 'the Amazon.com of professional services.' The disruption it brought provoked diverse reactions across the firm, from outright skepticism and indignation, to excitement, and even evangelical fervor. While challenging 'the way we do business around here' caused rancor, it was TechFirm's re-engineered version of 'why we do business around here' that stimulated a firm-wide transformation. Left behind was the traditional push model of billable hours for services rendered. What had revolved around professional experience and long-held practices for clients had been eroded in the new digital age. Clients engaged with TechFirm in co-creating their own service experiences. Moreover, new experience domains were designed that only a few years earlier would have been thought of as impossible.

DEMOCRATIZING INNOVATION

In its relatively short life, TechFirm had, through experimentation, unknowingly sidestepped the propensity to reduce innovation to a mechanistic, replicable process. TechFirm's CEO had adopted many of the tools associated with design over a 30-year career in business, acquiring a 'maverick innovator' profile in the process. Yet when the conservative bastions of the firm embraced the TechFirm approach to designing and selling experiences, innovation capability became a reality.

Design thinking is not a linear, step-by-step process. Its value for business innovation erodes with systemization. As TechFirm showed, design works best when it can enable individuals and teams to become comfortable with understanding, creating, testing, and adjusting both their practices and the thinking that informs the practices. That is, design bolsters ambidexterity capabilities by encouraging the dynamic between certainty and uncertainty, and at a broader level, exploit and explore.

Design thinking facilitates ambidexterity capability by democratizing innovation across the parent company and beyond to key stakeholders, seeding capability from the inside-out and outside-in. For example, as TechFirm matured, it moved its innovation socialization process from Yammer to a cloud-based platform developed in collaboration with a university entrepreneurship and innovation program. The resulting Innovation Academy broadened the creative commons by including customers, government agencies, and educational institutions. It also sharpened the ideation process around themes aligned with the company's overall strategy and its eight core service arms. A broad range of social media technologies was integrated with the Academy as well. In fact, the Innovation Academy's success led to a new Leadership Academy, offered as a value adding service to the firm's client base, and was supported by eclectic, group-based Innovation Cafés engaging participants with diverse backgrounds, expertise, and positions.

Design thinking and its accompanying tools enabled TechFirm to create and sustain a fluid disequilibrium between exploit and explore, converting the products and services of today to the client experiences of tomorrow. Furthermore, they shifted the organizational mindset from having the right to innovate to having the responsibility to innovate. Consequently, TechFirm overcame the most serious challenge facing innovation initiatives: The migration of resources from the explore domain back to exploit when aspirations are not immediately achieved.

LESSONS FROM THE EDGES

TechFirm recognized the so-called opportunities at the edges, while understanding the need to innovate at the core. However, it was not just a matter of setting up an idea capture tool, letting everyone loose, and then bringing together cross-functional teams.

It might be tempting to draw parallels with Steinbeck's observation that, 'ideas are like rabbits, you get a couple and learn how to handle them, and pretty soon you have a dozen' (1947, p. 123). We caution that ideas cannot be scaled in a linear fashion. Reducing innovation to a lowest common denominator can have disastrous effects. Design thinking is best employed when it becomes embedded at all levels within an organization; ambidexterity capability springs to life, unconstrained by the shackles of convention and conservatism.

TechFirm, over its life cycle, escalated its use of design from experimentation to embeddedness. Where at the early stages of its existence it was viewed skeptically by the senior members of the parent firm, it achieved acknowledgment for its role in turning the fortunes of the entire company around. One result was that similar initiatives have recently been introduced in several other countries in which the firm operates. Key staff members from TechFirm were seconded to oversee the new business units. Although the core tenets of TechFirm's approach were adopted, the business models received some customization in order to align with the local cultural and business imperatives.

Given TechFirm's unique and instructional journey, we offer the following propositions for consideration. While not written in stone, they are the result of considered analysis of numerous cases, theory, and experience, although TechFirm's journey remains at the heart of our lessons. On a cautionary note, we emphasize that design-led innovation cannot be reduced to a formulaic activity. Innovation by nature is often messy, chaotic, and intangible. In addition, contrary to popular assumption, it is not for everyone!

Innovation does not happen in a vacuum. TechFirm pulled down the walls of secrecy, protectionism, and a reluctance to share by creating a business unit wherein such practices were anathema. They avoided becoming too precious about any idea, practice, and even client. TechFirm's CEO drove a culture reliant on an 'open-source' philosophy. The collaboration proved central to the success of the unit, especially because it placed the end-user at the center of everything. As a result, TechFirm rapidly established a breathtaking network of partners, collaborators, and supporters, all with a fierce appetite for doing things differently. By challenging the closed circuit mode of operations that existed, TechFirm liberated a torrent of creative thinking, that preceded and shaped significant new products and services. While the cross-pollination of the firm's latent knowledge with newfound innovation delivered some significant improvements, it was through agile execution that the firm yielded exponential returns.

Design thinking goes beyond just product development. Although design thinking initially proved useful in business for product design, its effectiveness goes well beyond producing things better. At the heart of a design paradigm—as TechFirm exemplified—is a process of discovery. It always seeks to locate then amplify the value and impact of innovation on the end-user. It is about engaging with the actors in the value chain so that more empathetic and effective outcomes are achieved. This is just as true for internally facing endeavors as it is for market-facing endeavors. For example, ambidexterity capability can be encouraged by understanding the needs of end-users from their perspectives; a way of looking at the world that tends to reveal how organizing elements such as work unit structures, systems, and processes are designed for internal rather than customer convenience. Placing end-users at the core of market-facing activities reduces the likelihood of delivering poor value to customers. Inculcating design thinking in the organizational psyche unlocks a firm's potential to both think and do things differently. It also challenges firms to ask the right questions, as many organizations waste tremendous resources interrogating and defining the wrong problems and opportunities. Finally, a design model underscores the importance of speed to market. Agility in execution across both exploit and explore domains provides fertile ground for seeding ambidextrous capabilities.

Spend time on defining the problem or opportunity. In most organizations, and especially in business, the default approach means solving the most immediate, obvious problems as quickly as possible. Dealing with the noisiest problems, however, often results in the use of off-the-shelf solutions based on past experience. If a resolution worked once before it tends to get readily recycled. However, our research with TechFirm showed that in many cases the perceived problem or opportunity was not the real problem or opportunity. TechFirm's CEO continually asked his team to interrogate 'what is?' so that a better 'what could be?' was co-designed with and delivered to the end-users. Straight to solution approaches tend to deliver outcomes that do not meet the needs of the end-user. In fact, most problems come about in the first place because managers are too hasty to assume that they know what will maximize customer value. Instead of an immediate reaction to a problem, issue, or opportunity, design thinking prioritizes an engagement with end-users, substituting a 'push' with a 'pull' approach. From the perspective of the end-user, a superior understanding of deeper issues can be exposed, leading to better long-term, value-enhancing solutions whether for employees, clients, customers, or suppliers and distributors.

Leaders at all levels have to walk the walk. Earlier forays into innovation by TechFirm's parent firm were top-down driven with mandates broadcast about what everyone was supposed to think and do. The problem was that these dictates were not reinforced by actions, and little energy was given to explaining why the changes were critical in the first place. Calls for innovation—supported by lackluster and vague prescriptions for action—were dutifully received but blithely forgotten before business as usual resumed. TechFirm, in contrast, worked to engage firm executives, senior partners, line managers, and team leaders by spanning siloed boundaries. Leaders were subsequently better able to see the value in exposing the previously dormant creativity that had covertly resided in the collective service arms of the business. Senior partners and line managers slowly recognized a new suite of opportunities through innovations to the current business. At the same time, new revenues from more radical innovations inspired excitement and optimism. By seeding innovation from the core—the exploit activities of the firm—TechFirm embedded ambidextrous capability across the entire firm. Although the transformation proved both challenging and traumatic, it produced undeniable returns. A $300 million business that was losing clients and staff at an alarming rate morphed into a $1.5 billion business, attracting bright, new talent in the process.

Scale your commitments. TechFirm eschewed detailed planning in favor of end-user engagement. Their 'low-fi' approach to concept testing and prototyping in partnership with clients reduced the risk of committing resources to expensive new products and services that turn out to be unwanted. By co-designing solutions from the outset, resources can best be leveraged when and where they are needed. As our research demonstrated, the probability of delivering more innovative and valuable solutions was proportionate to the intensity of customer participation in the development cycle.

Galvanize through design. TechFirm used design thinking, and its associated design tools, to cut across organizational boundaries, offering everyone in the firm, no matter what their status or area of expertise, a common language of expression. A common innovation language helped to translate the diffused explosion of creativity into a usable form. It also made teams from diverse parts of the firm realize that their respective problems were not necessarily all that different, and that a brilliant idea in one area could be modified and adopted in another. TechFirm's CEO talked about embedding design in the DNA of the firm. To him, this meant taking a far broader view than just product design. Rather, design thinking should underpin an organization's entire mindset and identity, the forerunner to ambidexterity capability.

Ensure that viability does not get overlooked or 'bolted on' at the very end. Our research with TechFirm exposed how easy it can be to get caught up in the creativity cycle and lose focus around delivery; that is, the 'wow' without the 'how.' TechFirm's winning formula relied heavily upon interrogating the issues emerging during the co-design process concerning desirability, feasibility, and viability. Execution and agility were at the forefront in the design process, ensuring that meaningful value for the end-users was always prioritized.

Measure, acknowledge, and reward appropriately. Despite its myriad successes, TechFirm struggled to get its measurement and reward structure right. The McKinsey 'Global Survey on Innovation and Commercialization' (2010) reported that more than 70% of corporate leaders highlighted innovation as a top-three business priority, but that only 22% actually set innovation performance measures. TechFirm's parent firm measured innovation on the basis of new revenue growth over a three-year cycle. Although any kind of innovation metric distinguished the firm from most others, the emphasis on revenue reflected an exploit logic rather than one more concordant with a dual paradigm.

TechFirm's CEO preferred to view innovation measurement in terms of 'value to the company and its employees.' He pointed out, for example, that efficiency gains from innovation enabled services to be delivered faster and cheaper with lower error rates. However, while the returns to the firm were measurable, they were also obscured by clunky lag-based metrics, ill-equipped to accurately capture the real costs and benefits of innovation. Likewise, the value-add to current business through service extension via digital delivery or complementary products was only ever partially captured. Not only that, TechFirm's CEO lamented the difficulty in measuring the value of new capability development and alternative career pathways. Most of the theoretical guidance remains mired in exploit thinking. Lockwood (2007), for example, proposed a four-category framework for evaluating the added value of design in business: (1) 'more profit'; (2) 'more brand equity'; (3) 'more innovation'; and (4) 'faster change.' Kaplan (2014) argued for innovation measurement 'to get real results,' focusing on hard metrics for profit and revenue, product versus services revenue split, market share, customer loyalty, and licensing and royalty income. However, he maintained that these hard metrics should be supplemented and balanced with softer metrics around behaviors relating to leadership, employees, and customers.

The key to measuring and rewarding innovation lies firstly with the intent to do so, then in finding the right things to measure by using a customized set of metrics aligning with strategy, goals, and desired behaviors. This may involve some experimentation, but that is after all part of the design approach.

CONCLUSION

Is design thinking a revolutionary development for business innovation or is it a 'failed experiment'? Results from the Design Management Institute's (DMI) 'Design Value Index' suggested that design-centric, innovation-driven companies such as Apple, Proctor & Gamble, SAP, and Nike, have realized returns of 211% over the S&P 500 average. In fact, the three key takeaways from the DMI's research were reflected in our own research into TechFirm and its parent company. First, with committed leadership and senior-level support, effective design-led capability can be fast-tracked. Second, design, when strategically orientated, can be transformational for service and consulting-based organizations. Third, co-creation puts the end-user firmly at the center of a problem or opportunity so that more optimal solutions are created from the outside-in.

Our data based on TechFirm's innovation life cycle reinforced the nonlinear nature of innovation. We also witnessed the failure of reductionistic attempts to commoditize design thinking, which tended to produce little more than a mechanistic management fad. However, we also saw how powerful design thinking and its associated methodologies and tools can be in democratizing innovation in a conservative, declining firm.

Given that an organization's human capital comprises its most valuable asset (and greatest cost), changing the way individuals and collectives think and act around complex issues, as well as the way they deal with uncertainty and ambiguity, represents no small achievement. A defining lesson from the TechFirm case was that ambidexterity capability accompanied a human-centered, fuzzy, and fluid approach to innovation. Herein lies an appealing aspect of the design thinking approach to ambidexterity capability; it is not just another linear, systemized model that prescriptively attempts to organize that which is probably better left as emergent. Rather, if individuals and firms are bold enough to experiment, then innovation can be taken from $N + 1$ to N^X.

References

Beckman, S. L., & Barry, M. (2007). Innovation as a learning process: Embedding design thinking. *California Management Review, 50*(1), 25–56.

Brown, T. (2009). *Change by design: How design thinking transforms organizations and inspires innovation.* New York: Harper Business.

Buchanan, R. (1992). Wicked problems in design thinking. *Design Issues, 8*(2), 5–21.

Hobday, M., Boddington, A., & Grantham, V. (2011). An innovation perspective on design: Part 1. *Design Issues, 27*(4), 5–15.

Kaplan, S. (2014). How to measure innovation (to get real results). *Fastcodesign,* January 10, 2017. https://www.fastcodesign.com/3031788/how-to-measure-innovation-to-get-real-results.

Lockwood, T. (2007). Design value: A framework for measurement. *Design Management Review, 18*(4), 90–100.

Lockwood, T. (2010). *Design thinking. Integrating innovation, customer experience and brand value.* New York: Allworth Press.

Martin, R. L. (2009). *The design of business: Why design thinking is the next competitive advantage.* Boston: Harvard Business Press.

McKinsey & Company. (2010). Innovation and commercialization, 2010: McKinsey Global Survey Results. January 17, 2017. http://www.mckinsey.com/business-functions/strategy-and-corporate-finance/our-insights/innovation-failure/-and-commercialization-2010-mckinsey-global-survey-results.

Nussbaum, B. (2011). Design thinking is a failed experiment. So what's next? *Fast Company,* January 3, 2017. http://www.fastcodesign.com/1663558/design-thinking-is-a-failed-experiment-so-whats-next.

Steinbeck, J. (1947). Interview with best selling author: John Steinbeck. *Cosmopolitan, 18,* 123–125.

Verganti, R. (2009). *Design-driven innovation: Changing the rules of competition by radically innovating what things mean.* Boston: Harvard Business Press.

CHAPTER 8

The Efficient Innovator

Abstract This chapter reviews the contents of the book and its central arguments. It acknowledges that external change operates as a consistent presence rather than as an exception, but that a productive organizational response incorporates both instability and stability. This culminates in a paradox of sorts, wherein sustainable success demands a commitment to simultaneously high levels of both innovation and control. It is at this energetic intersection of both where ambidexterity capability receives its impetus. The chapter highlights the book's proposition that success requires simultaneous exploration and exploitation in high doses, encouraging tension and discontinuity. Finally, the chapter summarizes the argument for a duality ecosystem fostering ambidexterity because it augments the conversion of ideas into commercial implementation without compromising speed.

Keywords Paradox · Duality ecosystem · Conversion · Discontinuity

INTRODUCTION

Perhaps the most revealing measure of an organization's performance lies with its ability to resolve paradox. For example, nearly two decades ago, Molinsky (1999) noted that organizational change depends upon management, but that management intervention frequently decreases the likelihood of change. Perhaps this is because most organizational change

© The Author(s) 2017
A.C.T. Smith et al., *Reinventing Innovation*,
DOI 10.1007/978-3-319-57213-0_8

attempts focus on new conditions (Salem 2002) usually measured in terms of more equilibrium, stability, and control. In this book, we argued for an organizational paradigm embracing the seemingly paradoxical pursuit of both innovation and control, or what has become known as the 'explore—exploit' dilemma.

We began this book by acknowledging that external change operates as a consistent presence rather than an exception. Our subsequent argument followed that a productive organizational response incorporates both instability and stability. This culminated in a case for a paradox of sorts, wherein sustainable success demands a commitment to simultaneously high levels of both innovation and control. It is at the energetic intersection of both where ambidexterity capability receives its impetus.

Organizations in Western democracies have traditionally embraced economic rationalism, an orthodoxy that prioritizes science, analysis, linear thinking, and purposive, ordered planning. Concepts such as sustainable futures, innovation, long-term growth, competitive collaboration, and people-focused management were not part of the rationalist lexicon. However, as organizational boundaries blurred into global markets, organizations realized that they had to manage 'loose-tight' relationships and establish structures that enhance flexibility and responsiveness, at the same time as bolstering performance efficiencies.

In this book, we proposed that success requires simultaneous exploration and exploitation in high doses, encouraging tension and discontinuity. By adopting what we called a 'dualities approach,' we argued that organizations can better wield their 'mutually enabling constituent' parts to enact change (Farjoun 2010, p. 205). In the final comments of this book, we explore the implications associated with embracing the dualities challenge. A productive way of looking at the challenge involves accepting and even intentionally escalating the tension between explore and exploit organizing forms. Organizational reality maps readily with the gray area that organizing tension produces. It comes in the form of interrelated dualities such as stability and change, control and innovation, discipline and freedom, individual autonomy, and collaborative teamwork. All of these micro-dualities help to create an environment where ambidexterity capabilities can flourish. Next, we review some of the key concepts introduced in this book, including uncertainty, paradox, and dualities. Later in the chapter we offer a conceptual illustration specifying the constituents of dualities that help navigate concomitant innovation and control.

Uncertainty for Innovation

Early in this book, we suggested that the relationship between new (flexible for innovation) and old (traditional for control) forms of organizing might be better understood in terms of complementary tensions rather than contradictory powers. These tensions have been variously labeled paradoxes (Quinn and Cameron 1988), dilemmas (Stace and Dunphy 2001), dialectics (Mittroff and Linstone 1993), competing goals and values (Cyert and March 1992), and dualities (Evans and Doz 1989; Pettigrew and Fenton 2000). Some subtle distinctions between these terms may be drawn (Sanchez-Runde and Pettigrew 2003): paradoxes concern contradictions that do not always need resolution; dilemmas represent either/or situations that need to be resolved one way or the other; and competing goals and values imply the need for opposing parties to lobby and negotiate over alternative courses of action. We think, like Sanchez-Runde and Pettigrew (2003), that dualities represent an amalgam of paradoxes, dilemmas, dialectics, and competing goals and values. Unlike the others, dualities represent forces that need not be equitably balanced because although they appear paradoxical or contradictory, they are in fact complementary. A dualities-aware approach, therefore, recognizes that both extremes have merit. Rather than seeking resolution toward one extreme, organizational decision making and action needs to encourage a constructive tension between the two poles (Evans 1999). This insight has formed the foundational premise for this book.

By implication, surviving and succeeding in the knowledge economy means that organizations need to harness the power of dualities by encouraging a constructive tension between the collision of continuity and change. From a duality perspective, continuity and change represent an overarching meta-duality, fundamental to building healthy organizations. Rather than favoring one extreme over the other, organizations should adopt a duality-sensitive mindset, which recognizes the merits of both sides of the duality continuum. As a result, they are better placed to meet the simultaneous challenges present in contemporary organizations: global/local, autonomy/control, flexibility/efficiency, centralization/decentralization, and order/disorder.

The consequences of ignoring, or not managing, both sides of the duality continuum are highlighted in Gordon's (2005) ethnographic study of one of the world's largest police organizations. Well-intentioned reforms aimed at creating a more democratic, empowered workplace failed because senior management did not recognize the nature and extent to which existing organizing structures, processes, and boundaries would impact on, and ultimately prevent the successful implementation of the new more innovative initiatives. Responding effectively in uncertain markets generally involves more rather than less direction from central headquarters. As we pointed out in the first chapter, 'forms of organizing' are evolving, not forms of organization, and these dynamic forms are typically embedded in larger, more stable, bureaucratic organizational structures. In short, abandoning control for flexibility does not work any better than giving up on innovation in favor of efficiency.

On a practical level, the dualities concept illustrates how tensions in forms of organizing are managed not through definitive resolution but through adjustments. The concept also reinforces the need to discard assumptions about opposing values, instead replacing them with an appreciation of the distinctive yet related qualities of organizing forms. For example, the change-preservation and order-flexibility tensions do not need to be interpreted as one-dimensional choices. Flexibility might be essential in a turbulent environment in order to find new paths to innovation, but order is also necessary to ensure that innovation is focused and relevant. In other words, new forms of organizing cannot operate as substitutes for traditional forms because the very structures in which they exist are boundary-laden and fundamentally bureaucratic. New forms of organizing, therefore, might be better viewed within the concept of dualities, which embraces the logic of optimization rather than resolution. An optimization response reflects some fundamental decisions to be made in the engine room of management (see Fig. 8.1). Dualities present a different paradigm, accommodating gray areas and illustrating how paradox can be productive in fostering an environment conducive to ambidexterity capability.

FROM PARADOX TO DUALITIES

Paradox has been described as a 'middle way' of thinking where opposites are seen as interdependent entities that together comprise a totality (Chen 2008; Eisenhardt 2000). We have argued that paradoxical thinking fuels the dualities mindset. It encourages diverse perspectives, accepts contradictory elements, and reveals multiple organizational realities in which both innovation and control feature. We have proposed that a dualities lens allows leaders and managers to explore and exploit. It compels sensitivity and receptiveness to the complexities, ambiguities, and contradictions intertwined in the day-to-day routines of organizational life.

The next question we asked was, how can organizations mediate between the simultaneous need for innovation *and* control? Duality theory suggests that the key lies in allowing, and even deliberately encouraging, a state of tension to emerge (Lewis 2000). It conceptualizes organizational performance as a process of simultaneous innovation and stability, incorporating complexity and contradiction without the need to remove, suppress, ignore, or deny tension or paradox. Duality thinking compels sensitivity and receptivity to the complexities, ambiguities, and contradictions embedded in day-to-day routines.

In our attempt to reinvent innovation—paradoxically by pursuing control at the same time—we examined how firms over the last decade and a half have experimented with flexible organizing forms and practices while maintaining their control systems (Pettigrew et al. 2003). For example, 15 years ago Palmer and Dunford (2002) showed that higher levels of formalization were associated with the introduction of new organizational practices. Moreover, the coordinating and direction-setting role of a corporate center emerged as equally important and relevant for firms operating in a complex and uncertain environment. Equally, slow bureaucracies introduced flexibility for innovation by engaging multiple, inconsistent, divergent, and even seemingly opposing factors. More recent work—as reviewed in Chaps. 2 and 3—came to view this flexibility as a form of dynamic tension comprising complementary, synergistic tendencies rather than contradictory forces. While varied in nomenclature, all these studies pointed to the imperative of managing for both change and stability at the same time.

Exploiting Paradox Through Dualities

Empirical work on organizing forms—like the cases we presented—demonstrate that high levels of bureaucracy can coexist with high levels of innovation. In fact, rather than a balance between stability and change, success lies with the tension that the two create when both are present in significant quantities. Contradiction can be utilized for performance (Clegg et al. 2002) when leaders and managers decipher how the dynamic tension generates complementary, synergistic tendencies, rather than contradictory forces. The concept map reproduced in Fig. 8.1 provides an illustrative example concerning the tensions and trade-offs between the dualities associated with innovation (explore) and control (exploit).

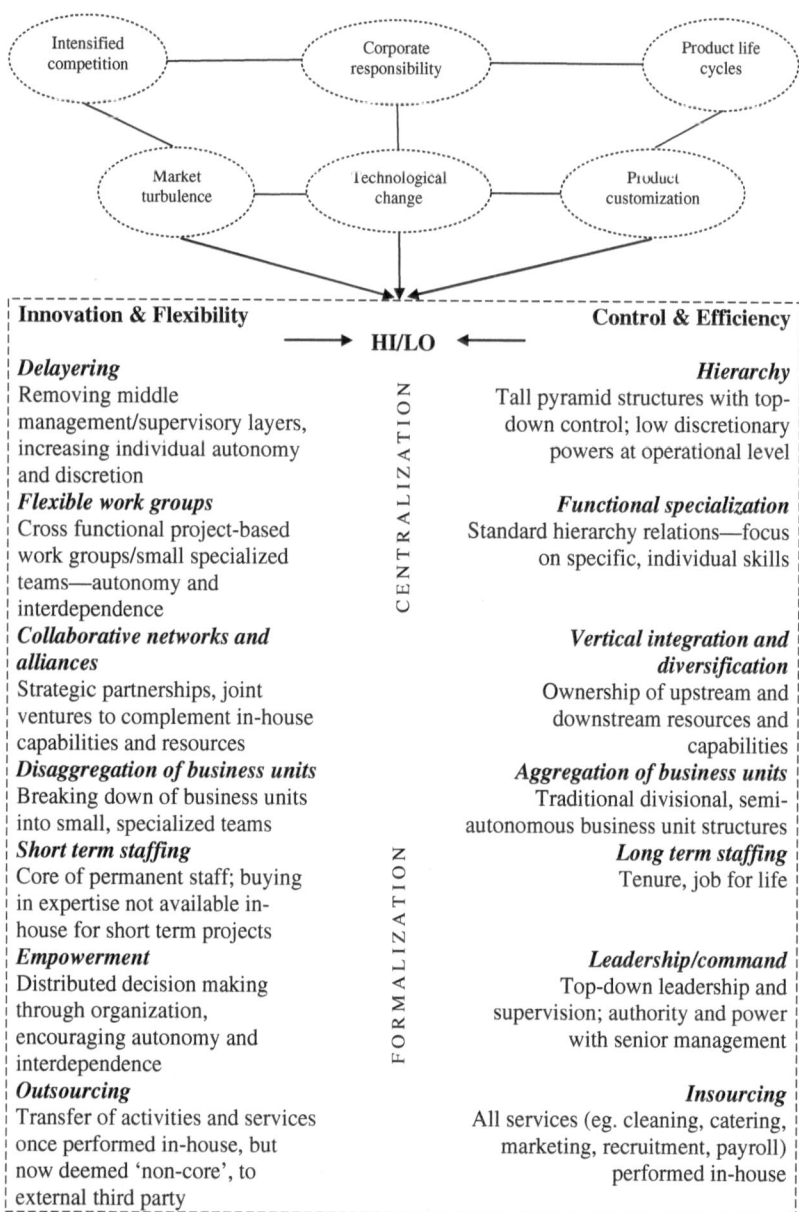

Fig. 8.1 Duality management framework

The model in Fig. 8.1 provides seven dimensions in which the innovation-control duality can be approached, each one representing the potential extremes of flexibility and efficiency. These are: (1) delayering/hierarchy; (2) flexible workgroups/functional specialization; (3) collaborative networks and alliances/vertical integration and diversification; (4) disaggregation of business units/aggregation of business units; (5) short-term staffing/long-term staffing; (6) empowerment/leadership and command; and (7) outsourcing/insourcing. These seven areas offer an organizing architecture around which ambidexterity capability can be scaffolded. Furthermore, these seven areas present specific management action targets in which ambidexterity capability can be measured.

He and Wong (2004) contended that the tensions between the two opposing poles need to be managed *continuously*, underling the dynamic, interactive nature of organizing dualities. A sufficient 'constructive' tension between each end of the continuum stimulates change and action. In the first dimension, for example, hierarchy provides structure and guidance, demarcating roles and responsibilities, as well as reporting relationships. But hierarchy should not compromise delayering principles that mitigate against overly 'tight' structures leading to rigidity, excessive reporting, and heavy-handed, inertial bureaucracy.

In the second dimension, functional specialization encourages professionalism and specialist expertise. When functional specialization creates tension with its complementary duality, flexible work groups, it lessens the risks of silo mentality and partisan behavior. A 'minimal threshold' of tension (Hedberg et al. 1976) between functional specialization and flexible work groups stimulates knowledge and expertise sharing and collaborative participation across teams and groups. Dualities such as delayering/hierarchy and flexible work groups/functional specialization also work by 'mutual specification,' conferring employees with high levels of autonomy. This freedom would be bounded by immutable principles (the 'tight' and focused component of hierarchy and functional specialization) that underpin their behavior and decision-making actions.

Dimension three, collaborative networks and alliances/vertical integration and diversification, and dimension four, outsourcing/insourcing, highlight the boundary shifts accompanying environmental turbulence characterized by transitions from economic to social, value-driven innovation. When organizations pursue excessive vertical integration and diversification, they risk weakening core business skills and competences.

Under a dualities lens, resolution or reconciliation between two mutually exclusive poles is no longer the end goal. Instead, focusing on the seven dual organizing dimensions exposes and facilitates the advantageous tension between organizing poles.

The second component in Fig. 8.1 recognizes six dynamics of the business environment including: (1) intensified competition; (2) product life cycles; (3) technological change; (4) market turbulence; (5) level of corporate responsibility; and (6) customization of products. We believe these six environmental characteristics would be valuable in analyzing the contextual variables influencing the extent to which certainty and uncertainty impact on the level of equilibrium along the dualities continuum. New and unimagined properties can emerge leading to new directions. Finally, the central column of Fig. 8.1 depicts 'formalization' and 'centralization' as integral components of the explore—exploit/innovation-control duality. However, it also captures the degree (HI/LO) to which organizational routines are formalized (i.e., bound by rules and protocols) and organizational decision making is centralized or decentralized.

INNOVATION REINVENTED

During the last two decades of the twentieth century, a proliferation of literature on new forms of organizing emerged proclaiming the imminent demise of bureaucracy. For knowledge-based companies in particular, it was argued that bureaucracy failed to make the best use of highly skilled knowledge workers or respond quickly and decisively to the demands of a technology-driven, customer-focused, and increasingly global marketplace. The arguments for new forms of organizing, or those embracing more flexibility and agility, were based on the view that environmental uncertainty is incompatible with traditional or bureaucratic organizational forms. Accordingly, network-driven, flat, permeable forms of organizing were seen to hold significant advantages over hierarchical, rule-centered bureaucracies. Despite these criticisms of bureaucracy and its trappings, there remained, however, many question marks over the evidence favoring the blanket replacement of old for new forms of organizing.

We now recognize that twenty-first-century leaders must confront a complex, turbulent, and global environment characterized by flux, but demanding constancy. Organizations face contradictory directives where long-term survival relies on embracing and exploiting tension. A dualities-aware perspective helps organizational leaders understand the fundamental interdependence and mutually enabling qualities of different approaches to organizing. As organizational boundaries blur and environmental turbulence becomes the norm, responsible leaders need to manage 'loose-tight' relationships by establishing structures that enhance flexibility and responsiveness within a strong risk and performance management framework. Organizational growth and long-term economic sustainability mean finding a balance between order and disorder through dualities (Farjoun 2010). In taking a further step towards enabling a dual organization, we argued for the development of ambidexterity capability.

We proposed five potential actions for building ambidexterity capability. As a result, dual exploit—explore tensions can be mobilized through: (1) critical integrative activities such as appointing skilled business unit leaders; (2) establishing critical communication and coordination linkages; (3) establishing appropriate resources and incentives for business unit leaders to support innovation and control systems and structures; (4) putting in place targeted human resources initiatives; and (5) introducing appropriate performance measures that recognize and reward both explore and exploit initiatives. Each of our cases offered detailed, contextualized examples of how these facets of ambidexterity capability were built in practice. They all displayed common characteristics revolving around the nonlinear nature of innovation and its unique relationship with the more predictable nature of mature business. We also saw how powerful design thinking and its associated methodologies and tools can be in democratizing innovation in a conservative environment. In all cases, ambidexterity capability accompanied a human-centered, fuzzy, and fluid approach to the innovation-control duality.

CONCLUSION

In this book, we have maintained that one pivotal way to take advantage of the explore—exploit tension involves the use of heterogeneous, agile communities capable of working in the gray area, where the normal rules and expectations become more elastic, and novel propositions can be tested in the real world quickly at a low cost and risk. Further, a duality-driven innovation-control program can lead to a 'new' business model where selling experiences takes priority over mechanistic service and product solutions. In order to make the offering successful, a firm must shift from 'and/or' approaches to reconciling exploit—explore tensions, instead optimizing both by explicitly affording everyone the right to innovate, while implicitly fostering everyone's responsibility to innovate. Vital to success is the commitment of a firm's leadership group to drive an ambidexterity capability mindset. The new mindset signals a commitment to rapid prototyping and concept proofs in the market combined with a design-oriented, user-based mode of thinking about client experiences. At the same time, lengthy commercialization plans and off-the-shelf service solutions are discarded.

We argued that a duality ecosystem fostering ambidexterity augments the conversion of ideas into commercial implementation, without compromising speed. Through such methods, expert and developing innovators fuel incremental innovation while consolidating innovation into a firm's conventional business units. A right to innovate becomes a core responsibility and highlights the centrality of driving the innovation culture deeper into the exploit landscape. The 'fluid innovation' observed in the cases demonstrated where the explore—exploit tension yielded an operational model. However, these findings do not fit as consistently with some interpretations of ambidexterity that advocate a dynamic oscillation between the two extremes, or of modular theories, which add new innovation structures and programs without a clear connection back into the firm's core, exploit business. As a result, we conclude by once again calling for the reinvention of innovation thinking, mobilized through ambidexterity capabilities. It is time to abandon the idea that core business and future business need to be balanced, or that either should be periodically prioritized. A dual organization chases both aggressively and uses the resulting tension as fuel to ignite ambidexterity capability.

REFERENCES

Chen, M.-J. (2008). Reconceptualizing the competition-cooperation relationship: A transparadox perspective. *Journal of Management Inquiry, 17*(4), 288–304.

Clegg, S. R., da Cunha, J. V., & e Cunha, M. P. (2002). Management paradoxes: A relational view. *Human Relations, 55*(5), 483–503.

Cyert, R. M., & March, J. G. (1992). *A behavioral theory of the firm* (2nd ed.). Oxford: Blackwell.

Eisenhardt, K. E. (2000). Paradox, spirals, ambivalence: The new language of change and pluralism. *Academy of Management Review, 25*(4), 703–705.

Evans, P. (1999). HRM on the edge: A duality perspective. *Organization, 6*(2), 325–338.

Evans, P., & Doz, Y. (1989). The dualistic organization. In P. Evans, Y. Doz, & A. Laurent (Eds.), *Human resource management in international firms: Change, globalization, innovation* (pp. 219–242). London, UK: Macmillan.

Farjoun, M. (2010). Beyond dualism: Stability and change as a duality. *Academy of Management Review, 35*(2), 202–225.

Gordon, R. D. (2005). An empirical investigation into the power behind empowerment. *Organization Management Journal, 2*(3), 144–165.

He, Z. L., & Wong, P. K. (2004). Exploration vs. exploitation: An empirical test of the ambidexterity hypothesis. *Organization Science, 15*(4), 481–494.

Hedberg, B., Nystrom, P., & Starbuck, W. H. (1976). Camping on seesaws: Prescriptions for a self designing organization. *Administrative Science Quarterly, 21*(1), 41–65.

Lewis, M. W. (2000). Exploring paradox: Toward a more comprehensive guide. *Academy of Management Review, 25*(4), 760–776.

Mitroff, I. I., & Linstone, H. A. (1993). *The unbounded mind: Breaking the chains of traditional business thinking.* Oxford: Oxford University Press.

Molinsky, A. L. (1999). Sanding down the edges paradoxical impediments to organizational change. *The Journal of Applied Behavioural Science, 35*(1), 8–24.

Palmer, I., & Dunford, R. (2002). Out with the old and in with the new? The relationship between traditional and new organizational practices. *International Journal of Organizational Analysis, 10*(3), 209–226.

Pettigrew, A. M., & Fenton, E. M. (2000). Complexities and dualities in innovative forms of organizing. In A. M. Pettigrew & E. M. Fenton (Eds.), *The innovative organization* (pp. 279–300). London: Sage.

Pettigrew, A. M., Whittington, R. L., Melin, L., Sanchez-Runde, C., Van Den Bosch, F. A. J., Ruigrok, W., et al. (2003). *Innovative forms of organizing.* London: Sage.

Quinn, R. E., & Cameron, K. S. (1988). *Paradox and transformation: Toward a theory of change in organization and management.* Cambridge, MA: Ballinger Publishing.

Salem, P. (2002). Assessment, change, and complexity. *Management Communication Quarterly, 15*(3), 442–450.

Sanchez-Runde, C. J., & Pettigrew, A. M. (2003). Managing dualities. In A. M. Pettigrew, R. L. Whittington, L. Melin, C. Sanchez-Runde, F. A. J. Van Den Bosch, W. Ruigrok, & T. Numagami (Eds.), *Innovative forms of organizing* (pp. 243–250). London, UK: Sage.

Stace, D., & Dunphy, D. (2001). *Beyond the boundaries: Leading and recreating the successful enterprise* (2nd ed.). Sydney: McGraw-Hill.

INDEX